Memory

Memory

FRAGMENTS OF A MODERN HISTORY

Alison Winter

The University of Chicago Press
Chicago and London

Alison Winter is associate professor of history at the University of Chicago.

The University of Chicago Press, Chicago 60637
The University of Chicago Press, Ltd., London
© 2012 by The University of Chicago
All rights reserved. Published 2012.
Printed in the United States of America

21 20 19 18 17 16 15 14 13 12   1 2 3 4 5

ISBN-13: 978-0-226-90258-6 (cloth)
ISBN-10: 0-226-90258-7 (cloth)

Library of Congress Cataloging-in-Publication Data

Winter, Alison, 1965–
   Memory: fragments of a modern history / Alison Winter.
      p. cm.
   Includes bibliographical references and index.
   ISBN-13: 978-0-226-90258-6 (cloth: alk. paper)
   ISBN-10: 0-226-90258-7 (cloth: alk. paper)   1. Memory.
2. Recollection (Psychology).   3. Memory—Research—
History.   4. Recollection (Psychology)—Research—History.
5. Psychology, Experimental—History.   6. Truth.   I. Title.
   BF371.W545 2011
   153.1′20904—dc23   2011017777

♾ This paper meets the requirements of ANSI/NISO Z39.48-1992
(Permanence of Paper).

To my mother, Judy Swartz—I remember *everything* . . .

CONTENTS

Acknowledgments ix
Introduction 1

1 On the Witness Stand 9
2 The Making of Truth Serum 33
3 Memories of War 53
4 Wilder Penfield and the Recording of Personal Experience 75
5 The Three Lives of Bridey Murphy 103
6 Securing Memory in the Cold War 125
7 Flashbulb Memories 157
8 The Law of Memory 179
9 Frederic Bartlett and the Social Psychology of Remembering 197
10 Making False Memory 225
11 Reliving and Revising the Past 257

Notes 273
Index 311

ACKNOWLEDGMENTS

I received a great deal of help in the course of writing this book, only a fraction of which I can acknowledge here. I would like to thank the following archives and libraries: the Wellcome Library for the History of Medicine, the National Archives and Records Administration, the National Library of Medicine, the Osler Library, McGill University Archives, Cambridge University Library, George Eastman House, Pueblo Historical Society and Archive, Harvard University Archives, the Boston Public Library, and the John Johnson Collection of Printed Ephemera at Oxford University.

I thank the following individuals for oral history interviews or information sent by e-mail, and in many cases for also providing me with bibliographical information and even original sources to help with my work: Ellen Bass, J. Kingston Cowart, Lynn Crook, Richard Epstein, William Feindel, John D. MacDonald, Brenda Milner, Alan Scheflin, John Smith, Bill Corcoran, Jennifer Freyd, Pamela Freyd, Peter Freyd, Deborah Glasser, Dick Grinker, John Kihlstrom, Jonah Klehs, Liz Lunbeck, Elizabeth Loftus, Emily Orne, Martin Reiser, Melissa Salten, David Spiegel, and Charles Tart. Many colleagues in history, science studies, and law gave me valuable advice and generous help of various kinds, including Albert Alschuler, Diana Barkan, David Bloor, Emily Buss, Lorraine Daston, John Forrester, Jan Goldstein, Tom Gunning, Bernard Harcourt, Dan Kevles, Ruth Leys, Eileen Groth Lyon, Andreas Mayer, Robert Richards, Simon Schaffer, Anne Secord, Rani Singh, Geoffrey Stone, and Alan Young. I have been very fortunate to have had several excellent research assistants who have greatly strengthened the research for this book. I particularly thank Mary Kate Stimmler and Jacob Betz for their imaginative, tenacious, and at times heroic work.

I have also greatly benefited from a publication grant from the National Library of Medicine, a fellowship from the Guggenheim Foundation, a Mellon New Directions Fellowship, and a grant from the National Science Foundation (SES-0551856).

I owe a great debt of gratitude to Christie Henry, my editor at the University of Chicago Press, and to the other staff of the Press for all their hard work, as well as to copy editor Alice Bennett.

Research that has contributed to this book has appeared earlier in the following publications: "Manchurian Candidates in the Cold War," *Grey Room*, 2011; "A Forensics of the Mind," *Isis* 98, no. 2 (June 2007): 332–43; "Film and the Transformation of Memory in Psychoanalysis, 1940–1960," *Science in Context* 19, no. 1 (2006): 111–36; "The Making of Truth Serum, 1920–1940," *Bulletin for the History of Medicine* 79, no. 3 (Fall 2005): 500–533; "Screening Selves: Sciences of Identity and Memory on Film, 1930–1960," *History of Psychology* 7, no. 4 (November 2004): 367–401; "The Chemistry of Truth and the Literature of Dystopia," in *Literature, Science, Psychoanalysis, 1830–1970: Essays in Honour of Gillian Beer*, ed. Helen Small and Trudi Tate (London: Oxford University Press, 2003).

Finally, I want to thank my family for their patience and support during the many years it has taken me to write this book, and particularly my children, David, Lizzie, Zoe, and Ben, who still believe, even now, that it will never, ever, ever actually be finished. Most of all I am grateful to Adrian, for everything.

INTRODUCTION

I am more intimately acquainted with my own memories than anyone else can possibly be. Am I, then, the best authority on them—and are you the best authority on yours? The answer may seem obvious, yet in today's world it is anything but. Most people, no matter their scientific background, agree that the stories we tell about our own pasts are central to how we understand ourselves. We have made them central to the operation of our society, too, in our institutions. Where we differ, sometimes violently, is on whether that status is warranted. Psychological experts claim that memories are capricious, error-prone, and partial—that they should be made the foundations for such important decisions, if at all, only after careful scrutiny.

In the modern era an extraordinary array of sciences has arisen to satisfy this scrutiny. We have become the first culture in history to subject memory—the basic recording mechanism on which both self and society rest—to the authority of the sciences. This book is about how that happened, and why it matters.

Memory examines the history of these sciences during the twentieth century. It focuses on experiments that explored the nature of autobiographical memories and on controversies over their reliability. It follows memory in action through psychiatric consulting rooms, surgical theaters, military battlegrounds, courtrooms, and self-help groups; through scientific papers, novels, diaries, scrapbooks, and films. The result is at once a social history of memory science, an intellectual history of remembering, and a cultural history of the mind.

The book ranges widely in time and subject, from early forensic techniques that aimed to improve the reliability of courtroom testimony to neurosurgical projects in the 1950s that aimed at obtaining conscious replays of memory records. The century it covers, culminating in the "memory wars" of the 1990s, represents a period of hyperbolic efforts in the recovery of memory.

There are many possible ways to tell the history of the memory sciences. One could, for example, focus on the handful of scientists whose work is most remembered and celebrated today. It would certainly seem a straightforward way to proceed. But in practice it would occlude many other projects that have sought to make sense of memory, some of them conducted outside the halls of the elite sciences. It would also give a false impression of the autonomy and self-sufficiency of the few favored figures. As this book's title suggests, the approach offered here is different. It offers a set of "fragments"— including some that would no longer appear in a currently compiled canon. On a practical level, a book on memory is perhaps inevitably going to be fragmentary given the unbounded scope of the subject. But there is more to it than that. These episodes offer opportunities to display the deep intellectual and cultural complexities of memory science in practice, and in its various times and places. Moreover, both the objects and the objectives of this book are best served by identifying discrete parts rather than by projecting in advance a coherent whole. We can best hope to understand broad historical developments precisely by bringing fragments of memory's history to life.

The historical questions that emerge from this approach echo in suggestive ways the psychological, clinical, and legal problems that dogged memory researchers throughout the twentieth century. They too often found their puzzles originating in a memory fragment. A patient lying on an operating table would suddenly begin to describe a few seconds of a musical concert she had attended years before; a chance reference to 9/11 would lead to a "flashbulb" memory of what the listener had been doing at the moment she learned that a plane had crashed into the World Trade Center; an adult in therapy would abruptly recall a particular moment that she then realized was part of years of sexual abuse. The fragment was always the starting point. Researchers took different tacks in a bid to make sense of it. The shape and texture of the history of modern memory science is, in the end, the result of their many and varied decisions.

As diverse as were the claims about memory, their references were often similar. Were autobiographical memories truthful evidence of past experiences or records of some other kind of personal information that bore a looser relation to the original event? If memories were "truthful," to what were they true? One could define truth by good faith (truthful remembering would be a good-faith statement of what one believed about the past); by fidelity to what one knew about a past event when it happened (for instance, the name of a friend in kindergarten or the flavor of the cake served at a long-ago birthday party); or by the preservation of the emotional experience of that past event (one's feelings toward that friend, or how much fun the party was). Different memory projects have put stake in different alignments of the relation, or distinction, between "truth" and "memory," based on different definitions of each. They also have alluded, usually implicitly, to different notions of authenticity. What would make a personal memory authentic, and what would make one think it false? If one could remember the emotional experience of the birthday party but was mistaken about certain factual details—the flavor of the cake, the names of the guests—was this more or less "authentic" than remembering the facts and mistaking or forgetting the emotional freight of the event? If a story about one's past, told in good faith, revealed something important and true about the rememberer's present "self," was this a kind of authenticity even if other evidence about the past flatly contradicted it? And might it be possible to develop techniques that could pronounce upon the authenticity of particular memories, or particular kinds of memories? The tensions between rival schemes played out in dramatic conflicts that grew more violent toward the end of the century. Whatever theories of remembering one subscribed to, everyone agreed that the stories we tell about our personal past define who we are in the present. The battles over the status of these stories amounted to a battle over the nature of the self.

Beyond memory itself, two themes run prominently through this book. One is media. The modern sciences of memory emerged when newly developed sciences of mind coincided with a proliferation of new media: technologies for recording, transmitting, and recreating sounds and images. Photography, the phonograph, and the moving image all developed between 1850 and 1900. They became identified with memory processes in a series of associations that shaped both how those processes were understood and how the technologies themselves would be used. Throughout the twentieth century, memory researchers continued to look to the most recent, cutting-edge recording technologies for insights into the nature of remembering. Modern accounts of memory have consistently been entwined with developments in recording and communications media. By focusing on this circumstance, *Memory* may tell us as much about how to approach the history of information media as it does about human memory.

There was never just one way to use recording technologies to think about memory. At one extreme, researchers suggested that they were models of memory itself. Perhaps, they reflected, memory was an internal recording that could be replayed at will. Central to these contentions was the idea that we could not only reflect on previous experiences, but actually reload them from some internal store and replay them in real time. The notion of reviving and even becoming deeply absorbed by past events was nothing new. What *was* distinctive was the creation of so many prominent and influential practices devoted to the idea of reviving past consciousness itself and reviewing old events, on the premise that they were stored as if on film or tape.

In places as diverse as the psychiatric consulting room, the surgical theater, and the cinema, one could find the personal past apparently being brought back to life, restaged or replayed. From the 1910s to the end of the century, a series of very different projects promoted the idea of memory as a stable archive or recording in the brain. The "truth serum" drugs of the 1910s and 1920s supposedly worked by suspending self-control, which was assumed to be all that was necessary to allow the "truth" to come out. A variety of communities would use these drugs over the years, some hoping for "truth," others for information (regardless of its truth status) that could be useful in a broader investigation, still others hoping for therapeutically significant experiences that might have no historical validity. Some of the same drugs were used in World War II to allow the free flow of blocked traumatic memories. In the 1950s, the electrode of a famed neurosurgeon seemed to work like the needle of a phonograph, playing past experiences in real time. Perhaps the most spectacular project of this kind was the famous case of Bridey Murphy, in which an amateur hypnotist helped his subject recover "memories" of an earlier life in nineteenth-century Ireland. News of the experiment electrified American society. Copycat experiments swept the country. At one time, one could apparently buy access to a previous "existence" at the going rate of twenty-five dollars.

Projects based on the idea of recording technologies thrived later in the century too—forensic psychologists dug for buried memories in the minds of traumatized victims, while academic psychologists and neurologists became fascinated by the apparent photographic richness of certain kinds of memories, some of which seemed to freeze a moment in time as if momentarily lit by a flashbulb. Autobiographical records in the brain were even institutionalized in legislation, with the passage of new laws that recognized the return of long-repressed memories by starting the clock on statutes of limitations at the time when the memories returned, rather than the original event.

But at the other extreme, researchers also used recording devices to define precisely what memory was *not*. For a number of scientists, the idea that memory is a recording device rests on an unrealistic fantasy of accuracy and permanence. Instead of practices that facilitated "reliving" a permanent record, they sought out ways to reveal an ineradicable role for interpretation—extending to "suggestion" and self-deceit—in the construction of knowledge and memory. Since the 1970s, researchers skeptical about the permanence of memory and the possibility of refreshing it have sought to explain away many vaunted revelations as the result of such suggestion.

Diverse concepts of media therefore became a leitmotif of the memory sciences. Another theme is the practice of law. Memory experts have often found it difficult to prove they have a better understanding of memory than ordinary people, and the question of the viability of witness evidence has been a key instance of this. Anglo-American courts rest on the expectation that juries can make good judgments about witnesses' memories by listening carefully to their testimony under cross-examination. This expectation is central to the law's authority in practice. But it was challenged repeatedly during the twentieth century. New sciences of memory were recruited in the name of jurisprudence and asked to provide foundations for the security of courtroom testimony. Psychologists' ambition was to place certain kinds of legal decisions on a surer and more accountable foundation. But the controversy over the new techniques produced the opposite result: it undermined the testimonial assumptions on which the science based itself.

Modern psychology turned out to offer a particularly problematic form of expertise, because its focus was that task so central to the jury's mandate: the evaluation of witnesses and their testimony. The problems with witnessing articulated by scientific experts would never be definitively dispatched by the methods of science alone. This past remains very much present in current debates over not only what kinds of experts are best to speak on questions of memory, but also whether expertise of any kind is appropriately interposed between witness testimony and the jury.

The courtroom is a key site for the sciences of memory because it is here that our society brings together individual memories (in the form of witness testimony), theories of what memory is, and anxieties about its accuracy. Contentions about expert witnesses, witness examination techniques, and the rules of evidence all focus on this conjunction. That is why a legal theme runs parallel to that of media through this history. It begins with the earliest disputes about eyewitness memory at the beginning of the twentieth century and extends beyond the memory wars at its end to very recent research on erasing or "dampening" memory.

When psychologists sought to bring sciences of memory into court, in one sense they offered to realize the ideal of a form of knowing basic to modern society. But in doing so they also questioned its foundations. As a result, central to modern disputes over the legal relevance of psychological techniques has been not the quality of the expertise but whether expertise is needed at all. If it is, then how can the principle of a lay jury be upheld? This issue is pertinent far beyond the courtroom. Everyone has a memory; we all have special, intensive knowledge of our own lives that others do not have. All of us, regardless of psychological or any other training, therefore have a degree of understanding of memory in our own cases that we could not claim with respect to many or perhaps most other scientific matters. Is this kind of apparently privileged knowledge also to be cast in doubt? And if so, what alternative could be more secure and lasting, especially in a context of continuous scientific change?

The human sciences, and sciences of memory in particular, are characterized by a distinctive relationship between specialists and nonexperts—one of dynamic mutual engagement. The figures in this history were continually engaged with what ordinary people believed about memories. These beliefs at times proved inconvenient to an expert's project, since people trusted their own minds rather than scientific pronouncements. This provided an overwhelming obstacle for one early project aimed at establishing the new field of forensic psychology. At other times, patients or ordinary citizens could play an important role as collaborators. The renowned experimental psychologist Frederic Bartlett's first career-defining experiments are a case in point: they began when members of the public were invited in to take part in experiments during an open house at the Cambridge psychology department. In some cases, the psychological categories of nonexperts became the seeds of experts' theories. In the 1950s, a celebrated neurosurgeon's claims about memory not only developed from collaborative relationships with his patients, who remained conscious during brain surgery, but emerged from conversations with them. They persuaded him to adopt their categories and embrace their interpretations of psychic phenomena.

In such incidents one catches glimpses of something beyond lay involvement. One sees hints of a vibrant culture of lay psychological exploration, independent of the world of academic expertise and research. This culture seems to possess elaborate understandings of how the mind works. The intense vogue for "past life regression" that developed in America in the 1950s was the major occasion for making it visible. Believers claimed that hypnosis had helped a Colorado housewife recover memories of a previous life in early nineteenth-century Ireland. Thousands wrote to the hypnotist concerned. They did so not because he had piqued their interest in reincarnation, but because they had their own research to report. It amounted to a spectacular array of confident, if sometimes extraordinary, explorations of what the mind could be held to contain and reveal. But for the episode that triggered the outpouring, it might well have remained invisible.

In the postwar years, psychologists and other scientists would often feel frustrated by what seemed to them intellectually illegitimate fads, sparked by irresponsible researchers or outright frauds. A similar concern is sometimes voiced today. Perhaps responses other than frustration are possible. One may be getting glimpses of a culture of lay psychological reflection that has thrived without the influence of the kind of parapsychological Svengali who has tended to bring it to their scandalized attention. It has a continuous history that has never been acknowledged but that has made real contributions at key points in the history of their expert disciplines.

It is after all inevitable that most acts of remembering (and their circulation, interpretation, and impact) take place in "lay" cultural settings. For most of us, our understanding of our selves therefore depends on what happens in those settings. Psychologists and neuroscientists—and historians too—have all articulated, in different ways, the idea that our understanding of who we are is intimately tied to our reflection on past experiences and actions. This is true of societies, communities, and individuals. When scientists confirm our intuitive categories—for instance, by claiming that our memories are even more perfect than we had hoped—it is exciting and perhaps emboldening. The idea that the arc of a baseball flying through the air from a father to a son is perfectly preserved, decades later, reassures us that the fine texture of a past that made our present is indeed part of the person we have become. By the same token, when an expert asserts that our memories are not veridical records of past experience but are in some cases whole-cloth fabrications, this can provoke outrage and resistance. The assertion may come from a Freudian, holding that a remembered childhood event is a fantasy representing old wishes or fears; or it may come from a social or cognitive psychologist, maintaining that records of past experience have become woven together with present ones in a "confabulation." Whatever the source, the claim that a memory that feels detailed and convincing is in fact unreliable is likely to be profoundly disturbing. That unease, and the resistance it can inspire, is not necessarily evidence of cussed anti-intellectualism. Indeed, it is not necessarily radically separate from the science to which it often seems opposed. Modern memory science itself has been shaped by interactions between ambitious experts and its various publics sparked by this friction.

The reason this matters is that the peculiarities of specific situations—a middle-class living room in Colorado, a Los Angeles police department, a surgical theater in Montreal—put their stamp on particular remembering projects, with the result that radically different understandings of memory flourished in different contexts and among different communities. Although the memory sciences of today are thriving, perhaps more than ever before, those deep divisions persist among expert communities. There are some fields in which scientific expertise challenges popular assumptions about how the world works. Evolutionary biology and climatology are but two of the most prominent at present. Of them all, memory science ranks at the top, along with evolutionary biology, in the challenge it represents to what are taken to be moral essentials. In the end, the sciences of memory may well prove at least as consequential for the future of our society and our selves.

I

ON THE WITNESS STAND

In January 1906, Elizabeth Martha "Bessie" Hollister, a middle-class married woman in her twenties, went missing in a working-class neighborhood on the North Side of Chicago, now Lincoln Park. Her family circulated a description: brown hair, blue eyes, gray dress. But it was already too late. Her body was found the next morning on a pile of refuse, with wire wrapped around her neck. The next day, police had a suspect: a young carpenter named Richard Ivens, who had found the body in the vacant lot behind his family's workshop on Belden Avenue.

After several hours of questioning, Ivens signed a lengthy confession, recounting his actions and maintaining that he had acted alone.[1] It seemed an open-and-shut case, but after a few weeks he dramatically changed his tune and announced that his confession was false. Yet Ivens did not claim he had consciously fabricated it. The police had used hypnotic techniques, he said—techniques powerful enough that he himself had temporarily come to accept their assertions as his own memories.[2]

Experts readily came forward to agree that Ivens's remarkable allegation could indeed be true. Psychologists from Chicago, the East Coast, and even Europe sent in their opinions, pulling the case into an ongoing controversy about the malleability of memory and the nature of suggestion. One of the foremost of these was Hugo Münsterberg of Harvard. Münsterberg saw in the Ivens case an opportunity to launch a broader campaign to demonstrate the need for expert psychological evaluations of witness testimony in general. His efforts became part of an intensifying discussion about the application of scientific knowledge to social problems. But when it was proposed that science be applied to memory—something that seemed definitively accessible to everyone—this discussion threatened to become acrimonious.

False Confession, False Memory

In the weeks after the Hollister murder, newspapers wasted little ink pondering the killer's identity. They did puzzle over various inconsistencies—for instance, Hollister's gloves and muff were never found, nor could police account for her whereabouts during the hours before they claimed Ivens had targeted her. But there seemed no doubt of his involvement.

It is easy to see why people assumed Ivens's guilt. He initially confirmed his confession to his own father as well as to the police. It was only after several weeks that he recanted, claiming the confession was a "false" memory imposed on him during the police interrogation.

Could the court be made to believe in such a thing? A false *confession* was hard enough to prove, but it was at least imaginable. A false *memory*, meaning a good-faith confession that had no basis in experience—this was unheard of. How could one replace one's own memory with a story that belonged to someone else's crime, especially when this would lead to one's own execution? Yet this was the defense's argument.

The trial began on March 6. Bailiffs struggled to keep order as the twelve jurors were sworn in before a throng of spectators. The presenting of evidence began two days later. The confession was immediately the central issue. Should it be admitted? If so, how much weight should it receive? The officer who had initially questioned Ivens testified (for the prosecution) that he had come to police attention almost immediately because he had found the body. He seemed suspiciously reticent, unwilling to answer questions, and very nervous. After some hours of questioning—during which time, the officer claimed, he had neither threatened Ivens nor offered him leniency, Ivens gave a detailed and even expansive confession of how he had sexually assaulted Mrs. Hollister and then strangled her. The defense attorney asked the chief of police whether he had studied hypnosis: he had, though not to secure confessions. Nor, he affirmed when pressed, had he used more subtle forms of suggestive influence.[3]

The defense's own presentation began with Ivens himself. He stated, sounding a little stilted and rehearsed, that he had confessed "under the sweatbox influence of the police." He had been nervous; they had "had me

[someone with] so little voluntary attention, and whose ability to associate his old impressions with new ones is so limited that when in the suggestible [state] . . . the patient finds it easier to acquiesce in, rather than to refute, any statement made by the suggestor. . . . We found the hypnotic somnambules to be a distinct and unmistakable type. They are always dependants and seldom, if ever, exhibit executive ability, preferring to be directed by those around them.[10]

Christison soon sent his text to psychological experts around the country and in Europe, soliciting authoritative support for his interpretation.[11] Among the first to respond were the most powerful figures in psychology at this time: William James and Hugo Münsterberg, both at Harvard, and Max Meyer at the University of Missouri. James worked from introspection and observation, Münsterberg from precision laboratory work; Meyer was developing a protobehaviorist approach. Despite their differences, they were all deeply involved in promoting and portraying "psychology" to nonprofessionals.

James had already told Münsterberg privately that he thought Ivens's confession was "quite preposterous." But his public statement for Christison was more restrained: Ivens was "probably innocent." A reprieve was necessary, James said, "for thoroughly investigating mental condition." Meyer concluded, more bluntly, that Ivens's confession was "the direct or indirect outgrowth of injudicious suggestions, coming from the police officers, [and] received by Ivens during the abnormal mental state above mentioned." Almost all the others Christison approached confirmed that suggestion could produce a false memory, and most of them warned that the circumstances of Ivens's confession made this a likely instance. All called for medical and psychiatric evaluation and a new trial. Max Meyer went further and baldly declared that "the jury was incompetent for this case." Only two experts, whom Christison quoted but did not name, had not been convinced that there was a fatal problem with either the investigation or the jury's decision.

Münsterberg was the most intrepid of all in supporting Christison's campaign. The Harvard psychologist made a confident diagnosis from afar:

I feel sure that the so-called confessions of Ivens are untrue, and that he had nothing to do with the crime.

It is an interesting and yet rather clear case of dissociation and autosuggestion. It would probably need careful treatment to build up his dissociated mind and thus awaken in him again a clear memory of his real experiences.[12]

Through "dissociation," Münsterberg claimed, Ivens's original memory had been severed from consciousness, and a new (false) memory taken its place. Only when Ivens eventually tapped into his original memory was he able to recognize and discard the false one.[13]

Behind this apparently strong agreement were elements of distinction. Christison's was not the only theory of what could have happened to Ivens to be held by his supporters. A different one would soon be put forward by Münsterberg himself. For the Chicago psychiatrists, the key issue was how easily Ivens could be led to embrace a suggestion. It was a matter of what an individual could be made to agree to. But for Münsterberg the central issue was whether the content of a memory could be changed if an individual was placed in an altered state of mind. In his eyes, it was a question of the content of memory itself and how it could be manipulated.

The Afterlife of Ivens's Memories

All of Christison's correspondents were interested in a tangle of issues brought together by hypnosis, relating to subtle influence, subconscious thinking, attention, and memory. They also shared a powerful interest in promoting psychology outside the clinic and the laboratory, wishing to give the field a greater stature in public culture. Münsterberg, who is now considered the founder of applied psychology, had written that it was absurd to leave psychology languishing within laboratories, "without immediate relation to the life around us." Within a decade, he forecast, applied psychology would be ubiquitous. Ivens's case became an instrument in the campaign to achieve that end.[14]

By early June, a few days before Ivens's appeal went to the state supreme court, newspapers were printing quotations from the experts' letters. But the letters were read not as evidence that Ivens had been wrongly accused, or of the need for psychological expert witnesses, but instead as proof of psychologists' gullibility and frivolity. The press saw in them all the intolerable smugness and noblesse oblige of East Coast intellectuals: "Illinois has quite enough of people with an itching mania for attending to other people's business without importing impertinence from Massachusetts. This crime itself, no matter who may be the criminal, was one of the frightful fruits of a sickly paltering with the stern administration of law. We do not want any directions from Harvard University irresponsibles for paltering still further."[15]

It seems that the court agreed, given the shocking speed with which they dismissed his appeal. A hint of wavering had come from Hollister's husband, who said he was worried that the court seemed to be leaning toward granting a stay of execution because of worries about the legitimacy of the confession.[16] And Ivens's defense counsel did not try to use the cache of letters, apparently thinking them unnecessary. But the counsel was mistaken and the widower's worry unfounded. Ivens's appeal was denied on the twentieth. The city prepared for an execution.

Ivens was hanged two days later, on June 22. The actual execution was closed to the public, but thousands of people thronged downtown Chicago, crowding Michigan Avenue (then called Michigan Street) and "waiting for the undertaker's wagon to leave the jail yard."[17]

However swift and decisive the appeal and execution might have been, debate about Ivens's guilt—and about the idea of a false memory—did not diminish. James and Münsterberg both issued new statements in the days after the execution. Christison also continued his campaign, expanding his pamphlet into a book and founding a new criminological society. In the next few years, the possibility of false memory sparked an increasingly intense debate about the very nature of legal evidence.

Memory in the Courtroom

Hugo Münsterberg had emigrated to America in the 1890s, hoping to realize William James's ambition to develop a "brass-instruments" psychological laboratory at Harvard. Münsterberg, who had been trained by Wilhelm Wundt in Leipzig, was devoted to a laboratory model of psychological knowledge making, using precision instruments to take quantitative measurements of various psychological effects. He worked tirelessly to call public attention to the power of psychological laboratory techniques, both at Harvard and in his popular writings.

He also sought wide social applications for his techniques. He was an architect of the various fields that collectively became known as "applied psychology," which integrated scientific psychology into education, industry, business, and, after Ivens, the law.[18]

Münsterberg was predisposed to embrace an opportunity like the one Christison offered. And for Münsterberg, hypnosis was not an exotic curiosity but a familiar tool in his work from as early as 1893, shortly after his arrival in America, when a clockwork hypnosis instrument featured prominently among his laboratory apparatus.[19]

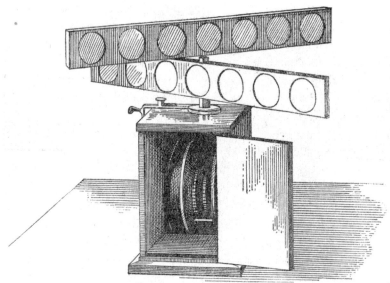

FIG. 39.—Mirror-hypnotizer.

"Mirror hypnotizer" or "hypnoscope," a device for creating a hypnotic trance in the psychological laboratory. From A. Macdonald, *Experimental Study of Children* (1898), 1167.

Sure enough, when he read about Ivens, he was sure that Ivens's memory must have been altered, and in some significantly diagnosable way. The alibi evidence seemed convincing, and the confession, by contrast, came across as "absurd and contradictory . . . like the involuntary elaboration of a suggestion." He soon discerned signs that Ivens was constitutionally vulnerable to dissociative states, in which the mind could develop "illusory memory as to its own doings in the past." But constitutional weakness alone could not produce a false memory, he noted, nor could aggressive interrogation. So some special event—a "sudden external shock or overwhelming fascination"— must have set the stage for such a rare phenomenon. But what could have produced such a shock?

Useful clues came from another puzzling memory case. Münsterberg had once seen a patient who came to him because her memory and personality had suddenly changed. In talking with her, he learned that on a recent occasion she had been tired and nervous, and by chance an intense light had been flashed in her eyes. This flash had brought about a dissociated mental state: "From that moment on her consciousness was split and her remaining half-personality developed a pseudo-memory of its own." He knew of another case involving a flash of light leading to a trance and an altered memory:

> She had gone to church in a condition of hopeless despair. The church was empty and, as she communed with herself, her hopelessness deepened. Then her eyes became fixed upon one of the shining brass lamps in the church, and of a sudden all was changed. She went into a trance-like state in which many disconnected memories of her early life and of happy times rushed to her consciousness, each accompanied by emotion, and these long-forgotten emotions of happiness persisted.[20]

This was a very different kind of explanation from the one offered by the Chicago psychologists, but it fit the work of a man who used light reflected off mirrors to produce altered states of mind.

Münsterberg suggested that something like this had happened in Ivens's case. There must have been some kind of "optical captivation of attention" as from an "eye-glass or the shining of the brass lamp." Nothing of the sort had been mentioned at trial, but Ivens himself supplied an answer just days before the execution. His normal memory seemed to return, and he described the terrifying sight of a revolver pointed at him during the interrogation. He especially mentioned the "flash of steel" as light reflected off the gun.[21] Münsterberg had found his flash. He asserted that Ivens's memory fit "the clinical experiences of the nervous physician and the mental experiences of the psychologist." Ivens could not possibly have designed such a perfect piece of information deliberately. The "flash of steel" followed by the "blur" and a "blank" were, for Münsterberg, the "missing link" proving his innocence.

For Münsterberg, the Ivens case was an extreme example of a universal vulnerability. Every human being, on this account, was a few short steps away from dramatic memory derangements, and some were already harboring multiple personalities. A chance experience, something as simple as stumbling in the street or glancing up at a building at a particular angle, could transform one's mental life.

There was a major problem with this claim. Münsterberg knew that many people were suspicious of psychological expertise in this sphere, regarding it as "another way of possibly cheating justice," of "emasculating" court procedure and "discouraging and disgusting every faithful officer of the law." Psychology had come of age in the past few decades, but lawyers did not yet see this. Even handwriting experts (whose skills Münsterberg derided) would be more welcome in court than a psychologist who wanted to evaluate witnesses' memory, perception, attention, and suggestibility. Moreover, there were already long-established legal conventions in place, aligned with common sense, for evaluating the credibility of testimony. The oath witnesses took, for instance, suggested that the alternative to truthful testimony was deliberate falsehood.[22] The new psychologists threatened this comforting idea. But the discomfort was worth it, Münsterberg insisted, in order to gain the benefits of the "magnifying glass" psychologists focused on mental processes. It was not sufficient to rely on the canniness of investigators and litigators to detect forgetfulness or lies—or manipulated memories. In a series of reflections that a century later seem oddly prescient, Münsterberg warned that heavy-handed pressure from police and lawyers was as likely to produce false statements as true ones—or even more likely. Reaction-time techniques, skillfully applied, could produce modern scientific certainty in unprecedented ways.

Münsterberg's arguments were much discussed by enthusiasts and detractors alike. One response in particular helped his cause: a play called *The Third Degree*, by the celebrated playwright Charles Klein, known for plots that were ripped from the headlines. The play was the third in a kind of American trilogy. Klein's first drama of the sequence, *The Music Master* (1904), focused on European immigrants struggling to make lives for themselves in turn-of-the-century America; his second, *The Lion and the Mouse* (1905), featured a young woman taking on a powerful business tycoon. Then came *The Third Degree* (1908). Klein intended it to be "instructive, by dealing with a subject that was of consequence to the welfare of the American people."[23] Klein had long been fascinated by suggestion, which he saw as the key to understanding and regulating the power of the theater itself. In 1905 he told the *New York Times* that the "mental images" produced by theatrical productions created powerful "auto-suggestions" in the minds of spectators. Once these mechanisms became better understood through the work of "physicians, metaphysicians and psychologists," it would become possible to regulate the theater "intelligently." In the case of *The Third Degree*, the theme of autosuggestion was also central to the play's content: Münsterberg's articles had convinced Klein of the danger of "hypnotic suggestion and visual concentration in police inquiry."[24]

In the play, a young man named Jeffries is interrogated by a policeman who forcefully repeats a story he wants the man to confess to. As he places a pistol on the table, he states, "You shot him with this pistol." The suspect winces at a flash of light reflected off the barrel. The stage direction instructs, *"His eyes are rivetted on it until his face assumes a vacant stare. Scientifically, this accomplishes the act of hypnotism and he comes under the influence of the will directing his will—he is now completely receptive."* The suspect, eyes locked on the pistol, repeats the confession.[25] A psychiatrist finds the confession "absurd and contradictory," the "involuntary elaboration of a suggestion put into his mind ... the law ought to recognize these scientific facts."[26] The lawyer's response? "The law doesn't recognize metaphysics and I'm afraid it never will until our lawmakers study science as well as politics." Eventually, other witnesses reveal that the death was a suicide, exonerating Jeffries.

Klein's reference to the gun and the light pointed to Münsterberg's warnings about the frailty of memory in general, and to his explanation of the Ivens case in particular. The play advanced a specific theory of *how* memory could go wrong. It endorsed Münsterberg's idea of a reflexive response to a visual stimulus (the light reflecting off the gun), producing a suggestible *state of mind*, rather than the "persuadable" states described by people like Parkyn and Christison, in which individuals became more willing to repeat what was "suggested" to them, but without a change in cognition or perception.

The play finished a two-year run in Chicago and New York without a single complaint about its central plot device.[27] It played in several other major cities over three years, bringing with it the Münsterberg theory. In New York, for instance, it caused an (unspecified) "upheaval" in the New York Police Department.[28] Reviewers reminded readers of Münsterberg's warnings

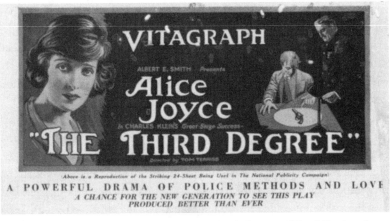

Advertisement for Vitagraph's film *The Third Degree* (1919). Courtesy of Derek Boothroyd.

and explained the psychological theory behind the title of the play.[29] The play was also dramatized as a film twice in the next two decades, first in 1919 from the Vitagraph Company, and then in 1926 from Paramount. Little is known about the first film, but the second was well received. Reviewers, who particularly liked the interrogation scene, treated as common knowledge the idea that such interrogation could produce a "hypnotic strain."[30]

Ivens's case was the first of several experiences that challenged Münsterberg's assumptions about memory. Later that summer he returned home from vacation to find signs of forced entry: glass and crockery were broken, furniture was overturned, and items were missing. Eventually, several men were arrested. At trial, Münsterberg testified that the burglars had entered through a cellar window. He said that he had "found drops of candle wax on the second floor," indicating they had entered the house at night, and that they left a large clock wrapped in packing paper on the dining room table.

The burglars might have felt unlucky in their victim: Münsterberg was a celebrity academic, renowned both for his knowledge of the mind and for his powerful intellect. He prided himself on (among many things) his prodigious memory. But to his chagrin, he soon found that he was wrong on most counts. The burglars had "broken the lock of the cellar door," not the window. A clock had been moved, but it was wrapped in a tablecloth, not in paper. There was candle wax, but it was in the attic, not on the second floor. Münsterberg counted up with alarm "how many illusions had come in."

Münsterberg decided that his mistakes could be explained psychologically. Elements of information that had struck him vividly had been retained particularly well—like the fact that there were candle drippings at all. But because he was disoriented and upset, his attention was less focused than usual on other information, and less of it was retained (for instance, the exact location of the wax).[31] The "illusions" came sometimes from his own imagination, which tried to fill the gaps in his memory, and sometimes from the police, who asked suggestive questions based on false assumptions (for instance, that the burglars had entered the house through a cellar window). Münsterberg became convinced that in the ordinary course of things—and not just in the extraordinary dissociative situation of someone like Ivens—our "perception" of events was "full of little blanks which our imaginative memory fills all the time with fitting associations." We were not usually aware of doing this. But techniques like his could reveal them, acting like "mental Roentgen [X-]rays which illumine the internal happenings."[32]

Münsterberg thought that anyone could become convinced of the power of suggestions to embroider memories, by conducting simple experiments. He carried out some of them himself. For instance, he showed a picture of a farmhouse room to a group of children and adults. Everyone examined it closely, and then they were asked what they could remember. The first questions were simple and "objective," such as the number of people in the room. Then, however, came more suggestive questions—the experimenter might ask if there was a stove when there had been none. Answers to these questions contained more errors as people "invented" suggested objects. The more suggestive the questions were, the more readily people elaborated their inventions.

To people with no psychological training, suggestion might seem a slippery and mysterious concept, but Münsterberg declared that it was part of ordinary brain functioning. Millions of neural paths led to the brain, he noted, and millions of paths led out again. He likened the cortex to a railroad or telephone system—a "great automatic switch-board." Every idea had a pathway, so when "excitement" was triggered in one pathway, the neural track to the opposite action was blocked.[33] Ideas remained vivid when the discharge channels remained open and became inhibited when they were closed. Suggestions kept these channels open and closed alternative paths. This process was largely unconscious, however, and neither observable nor controllable by the individual concerned. It took experimental expertise to detect and trace it.

Autobiographical memory clearly called for expert attention—not just in the laboratory, but in everyday life, wherever it was important to get the facts right. This was all the more urgent because people were so confident in their own memories. Psychologists, he came to believe, could teach people not to trust indiscriminately in their own memories and could help them decide how much value to place in a particular memory. He set out to promote this idea, beginning with another high-profile murder case—one in which he would find himself at odds first with the jury, and then with the press.

In the spring of 1907, America was riveted by the sensational trial of a union boss, William "Big Bill" Haywood. Haywood was accused of ordering the assassination of the former governor of Idaho, Frank Steunenberg. The murder was the culmination of years of violence in a struggle between a miners' union and the owners' association. A man named Harry Orchard confessed but said he had not acted alone. Orchard, who claimed that a conversion to

STUDYING THE EFFECTS OF COLORS ON JUDGMENTS OF TIME.

Experiments on perception in Hugo Münsterberg's laboratory at Harvard University. *McClure's Magazine* 1, no. 1 (1893).

Christianity had driven him to clear his conscience, gave a statement implicating Haywood, not only in the Steunenberg murder but also in seventeen others. During the trial in May 1907, before a standing-room-only crowd, Orchard recounted a series of assassination attempts and then the Steunenberg murder.[34] On direct orders from Haywood, he claimed, he had set bombs targeting police and strikebreakers.

Was Orchard telling the truth? *McClure's Magazine* hoped science could produce an answer. The magazine had followed Münsterberg's work since as early as 1893, when they helped him promote the gleaming new instruments in his laboratory.[35] Over the years, *McClure's* had followed his progress and the increasing power of his instruments to unlock the nature of mental processes. The reach of these methods had grown so great, Münsterberg himself claimed, that "there is now hardly a corner of mental life into which experimental psychology has not thrown its searchlight." Memory was the first dark corner.

McClure's asked Münsterberg to examine Orchard and write about his conclusions. Münsterberg accepted enthusiastically. This was an opportunity to show what his techniques could do. He packed his instruments in big wooden boxes, loaded them on a train, and traveled to Idaho.

Münsterberg arrived at the courthouse early enough to see part of Orchard's testimony. At first he was dismayed by the man's unprepossessing physiognomy: his profile was "most brutal and vulgar." His ear was deformed, and his eyes moved in a distressing, irregular way. It was easy to imagine him a murderer, harder to see him as the "sincere religious convert" that he claimed to be.[36] But Münsterberg was determined not to trust casual impressions, even his own. He needed to make a series of experiments on Orchard—reaction-time tests, to be precise. They alone could offer virtual certainty, unlike the kind of intuitive evaluation that was all he—and anyone else, including members of the jury—could make in the courtroom.

To carry out this test, the researcher read aloud a list of words. The subject was to reply with whatever word came to mind. At the beginning, the test words were innocuous. But soon the tester would begin to throw in words he expected to have emotional significance—in this case, "revolver," confession," and "murder." Test subjects were encouraged to assume that the test focused on their choice of words. But herein lay the trick, and the key to the experiment. In fact it was the elapsed time before the reply—measured to hundredths of a second and recorded for each response—that was significant. If the test subject had something to hide, he would censor the first words that came to mind, because these could reveal his guilt. His eventual innocuous response would take longer because his first idea had to be rejected.

Orchard was happy to take the test—a response Münsterberg initially dismissed as the confidence of a good poker player. Münsterberg told Orchard he would call out a series of words, and when responses came to mind, Orchard should "let them go without any reflection." Münsterberg also told him (deceptively) that he would learn everything he needed from the ideas that Orchard would "bring up."

The test began. At first, Orchard was consistent: his replies came after seven- or eight-tenths of a second. These first replies could be treated as a calibration of his natural response rate, because they responded to "neutral" words with no associations to the murder. Then Münsterberg read out words that did carry such associations. Orchard's responses continued to stream in at the same rate. His steadiness convinced Münsterberg that he had nothing to hide.

Münsterberg was satisfied. He packed his instruments and boarded a train back to Boston, where he planned to publish a pamphlet and articles after the trial was over. But toward the end of the journey his plans were derailed by a chance encounter with a reporter, who persuaded him to give an interview. He expressed in some detail his confidence in Orchard's truthfulness and Haywood's guilt. It was an indulgence he came to regret. The published interview triggered a wave of public opprobrium, particularly after Haywood was acquitted, putting the jury in conflict with Münsterberg. He became the object of contemptuous wit in the press, which sarcastically announced his invention of a foolproof lie-detection machine with which he had attempted to influence the course of the trial.[37]

Humiliated, Münsterberg canceled his plans for publication. When he eventually decided to write about the case, he avoided mentioning the conflict with the verdict and claimed that Orchard had revised his own memories to produce false statements about Haywood. His tests could still be counted a success on this account, because what they had evaluated was subjective veracity—good faith—rather than the ultimate truth or falsity of what was said. Assuming that the jury's decision was correct, this led to the inference that Orchard's "mind was diseased." He *must* have suffered from false memory.

The notion of false memory thus gave Münsterberg a way of reconciling his association tests with a jury's evaluation of the facts of the case. Writ large, the concept provided a way of trying (if a trifle desperately) to reconcile his program with the authority of juries generally. Paradoxically, to save both one needed to make what they had in common—memory—mean two things at once.

The Truth That Justice Demands

By spring 1907, Münsterberg looked back on a year of spectacular failure. He had failed to secure first an acquittal, then a reprieve, for Ivens. His own memory had let him down in the burglary case. And then the Haywood jury had flatly rebuffed him.

But Münsterberg's career was characterized by nothing if not his immense confidence. In that light these failures were opportunities, demonstrating the need to warn the world about the pervasive way that errors could creep into what the jury heard and credited. Testimony—even his own—needed to be as much the preserve of scientific expertise as blood or fingerprints. Just as a juror could not himself determine the origin of a blood sample, so he should not be expected to decide the credibility of a witness's account of his observation and actions.

So Münsterberg undertook a series of essays on psychology and testimony, chronicling his experiences and his analysis in high-profile magazine articles and assembling them for republication in a book called *On the Wit-*

ness Stand: Essays on Psychology and Crime (1908). Here he made most strongly the case that a psychological detective could bring out information beyond both lay understanding and ordinary medical expertise. Even doctors could take a "naïve view of psychical life." Psychologists had to be brave enough to confront "popular prejudice." Münsterberg had not initially reckoned with how painful and difficult this task would be. But he now felt that he was winning—that people in Chicago had been convinced by his view of the Ivens case and, more generally, that the press was beginning to report his forensic claims with more deference.

The time was therefore coming, Münsterberg argued, when these methods had to be accepted in court. As things stood, cross-examinations were often nothing more than demonstrations of the findings of psychologists like him, that "there is nothing more suggestive for some persons than a skillful question. Their influence could set in long before [opposing counsel could object to a] 'leading question.'" Just as the "bodily facts" in a case had to be examined by a chemist or physiologist, the "mental facts must be examined also, not by the layman, but by the scientific psychologist, with the training of a psychological laboratory."[38] Experimental psychology was a vital part of the process of leaving brutality behind, while making the search for truth more precise. The world would soon recognize that his views were in keeping with the "spirit of the twentieth century."

Mind and Media

It is worth taking a moment to consider Münsterberg's reflections about film. Analogies between film and memory were to become almost irresistible later in the century, particularly whenever the reliability of memory was at issue. And Münsterberg's *The Photoplay: A Psychological Study* (1916) was widely influential in his day, becoming a classic in the history of film studies. Just as forensic psychologists see Münsterberg as a founder of their field, so film studies scholars claim him as the progenitor of theirs. Münsterberg's account recalled themes in his forensic work. One might therefore assume that when later forensic writers made reference to film, their claims would echo his. In fact, postwar writers on memory did discuss film, but their claims were diametrically opposed to Münsterberg's photoplay argument. So it is worth taking a closer look at Münsterberg's psychology of film, if only to understand the profound shift that would take place when later writers enlisted film in their own writings.

In the years after *On the Witness Stand*, Münsterberg wrote many works that expanded psychology's reach into new areas—among them industry, business, and education. He also published a general psychology textbook with an overview of these new "applied" fields.[39] Then, in 1914, he was asked to write an essay on the psychology of film. Up to that time he had dismissed the entire medium as crass and uncultured. But as soon as he actually viewed a film, he became fascinated with the new medium and saw its potential as transformative. He came to think that his account of film applied to all aspects of mental life—including, prominently but not exclusively, memory.

Film, Münsterberg reasoned, had introduced new techniques to control a spectator's attention, perceptions, and emotions. It did this by externalizing the internal workings of the mind. It was as if a film were an embodiment of mental processes, but located in an external technology. If one were to imagine the film as the work of an individual, that individual's intimate thoughts and memories could be shared with someone else (the viewer) more directly and powerfully than through stage or the page, because of the way images could be assembled, delivered in close-up, and presented. Münsterberg specifically claimed that film could produce a sense of the past more powerfully than stage productions. It could juxtapose scenes from "now" and "then," either as a character's memory or as a straight recounting of past events. Film created the experience of memory, therefore, as well as memories themselves.

About the time of his photoplay essay, Paramount invited Münsterberg to design a set of educational sequences that would appear in the *Paramount-Bray Pictograph*, a "magazine of the screen" that combined educational newsreel sequences, animation, and travelogues. Advertisements touted the pictographs as a new way to "have people think" by invoking the "visualization of thought." Paramount promised no less than a "virtual motion picture university."[40]

By January 1916, Paramount had a number of "scenarios" for individual "pictographs."[41] These were not merely pedagogical narratives but experiments: cinema audiences subjected themselves to the pictograph "instruments." Spectators would learn about the mind by watching the film, not only because it would teach them about psychology, but because the film itself was an experiential event with potentially life-altering significance. Münsterberg described the series, titled "Testing Your Mind," as "demonstrations of psychological tests by which individuals can examine their fitness for special vocations." A table of contents in a newspaper advertisement offered a list of questions for viewers: "Have you a constructive imagination? Are you a square peg in a round hole? Can you suppress ideas? Are you suggestible? Are you mentally fitted for your work? Does your mind react quickly? Can your mind work quickly?"[42]

By May, audiences viewed the first pictographs. George Meeker, Paramount's editor of newspictures, told Münsterberg that "Does Your Mind Work Quickly?" and "Constructive Imagination" were the most popular.[43] These two, the only ones whose titles have survived, suggest that Münsterberg and Paramount were using the pictographs to stage educational experiments that also functioned as motivational or self-help projects. The first pictographs were so successful that Paramount asked Münsterberg for more so they could "enlarge" on the series that had already been screened.

In May, Paramount screened its pictographs in Chicago at an exposition convened by the national theater owners' organization. Münsterberg, as associate editor of the pictographs, gave a speech. He argued that the screen was a more powerful platform for psychology than any laboratory and that film could become a powerful vehicle for social engineering. "Much is heard about conserving the natural resources in this country," he announced, but relatively little was said about "conserving the natural talents of young people, about directing their ability in the proper direction." Pictographs could play a role here, helping to "direct a vast wealth of talent into their proper channels." Münsterberg's mental tests would allow people to identify "what characteristics equip [them] for special kinds of work," and to subject themselves to tests that would place them in different categories.[44]

Constructive imagination, for instance, was a quality that everyone had. Indeed, this was the same kind of attribute that Münsterberg had described (without giving it that name) in his accounts of memory change of 1906–8. But people who were stronger in this area had executive potential, because executives "must be able to grasp the meaning of a situation as soon as it presents itself." The test for "constructive imagination" screened a series of groups of letters, which viewers had to rearrange quickly in their minds to produce meaningful words. There were other tests to identify people who were good at noticing and retaining detailed information. Münsterberg concluded that anyone could benefit from his series, because it could identify the appropriate connection between the talents of any type of mind and its appropriate occupation. Motion pictures could be used as educational tools in any subject area, but he thought psychology was special; no other subject "can be more clearly presented through the camera." The kind of information he was actually delivering was vocational rather than—as one might have assumed from this description—an education in psychological theories, independent of their immediate application to the test subject.

Münsterberg's film project is interesting in light of his often-futile struggles to convince judges, lawyers, juries, and the general public of the legitimacy of expert authority in the face of conflicts with commonly held beliefs. Here, according to his own arguments, was a medium that could transfer the point of view of the film's author into the mind of a viewer. It did so by very rapid flashes of light that elicited a powerful neurophysiological response; so the famous image of the train barreling toward the audience was a counterpoint to Ivens's response to the flash of light on the gun barrel.

Eight years earlier, these experiments might have been designed to teach audiences that they could not trust their perceptions or their memories. Now, however, Münsterberg was preoccupied with vocational social engineering, a project that had elements similar to its influential contemporary, eugenics: identifying individuals with relatively stronger or weaker memories, perceptual skills of various kinds, and creative powers, so as to advise them on suitable careers. Many of the relevant characteristics, however, and some of the questions and experiments, were the same as in his earlier campaigns. How much of an individual's memory for details disappeared over time? After someone viewed an image that did not make sense, how did his imagination alter it to make it meaningful? How did individuals differ in their ability to perceive, remember, or imaginatively reconstruct information? Issues that in 1908 would have been framed around memory and the law were now to be defined in terms of the choice of appropriate career.

All this is worth bearing in mind, because the relation between film and mind suggested here is so different from what it would later become. Film was to play an important role in later accounts of memory, as people in a variety of circumstances sought to rethink that relation. But these had little connection to the account developed by the "father" of film studies, even when authors knew Münsterberg's psychological writings and saw themselves as emulating him. These later researchers represented film as a passive entity that took the impression of life experiences. Film was like the "wax tablet" of classical psychological writings, except that it did not soften and allow its impressions to fade over time.

Münsterberg had seen film very differently, as an active and transforming entity that was a way of intervening powerfully and subtly in mental processes. It was a lot like the mind itself, as Münsterberg portrayed it: a constructive influence that added to and restructured information. Indeed, it resembled Münsterberg's account of memory and remembering, in which stimuli and imagination conspired to alter the memory record, unknown to the rememberer. Because it bypassed the viewer's volition, it also erased the spectator's sense of its very constructedness.

April Fool

Although Münsterberg loudly proclaimed the success of his writings on forensic psychology, he was famously ineffective in his attempts to evangelize the courts. His core claim was that in principle *no* witness's memory could safely be evaluated by juries' common sense alone. He had no idea how difficult it would be to establish this claim.

John Henry Wigmore, professor of law at Northwestern University Law School and at this time the leading authority on evidence, led the response to Münsterberg's charges. He took umbrage at the idea that juries could not perform their most fundamental duty—evaluating a witness's credibility—without the help of a paid expert. Wigmore had already had a brief exchange of letters with Münsterberg after Münsterberg's articles appeared in magazines. In a 1908 supplement to his *Treatise on Evidence*, Wigmore referred readers to several psychological works by Münsterberg, and he invited Münsterberg himself to contribute an article to Wigmore's journal, the *Illinois Law Review*. Wigmore urged Münsterberg to see lawyers, not the general public, as the community he had to educate.[45] A year later, Wigmore wrote once more, warning that he was planning to "poke some fun" at Münsterberg's indictment of lawyers, a fair reply to "the fun you have had in publicly putting our profession in the pillory."[46] Wigmore's "fun" appeared in print in the April edition of the *Illinois Law Journal*. In a mock-trial transcript dated April 1, 1909, he subjected Münsterberg to a fictional suit for libel.

Wigmore convicted the defendant in *Cokestone vs. Münsterberg* of double standards. Münsterberg had claimed that although there was abundant scientific literature documenting the value of particular scientific techniques for evaluating witness testimony in court, lawyers chose to remain ignorant of it.[47] Wigmore examined psychological "witnesses" (quoting their published writings) to make the case that the literature did not in fact support the use of any particular psychological techniques in court. For instance, Wigmore quoted Freud's advice to law students: "In the laboratory experiments you will never be able to reproduce the identical situation of the real accused person." So although one should study people who had been accused of crimes, the results of that study should have no "influence on the decision of the magistrate." He also quoted legal writings to prove the authors were familiar with the psychological literature on memory and testimony. He even wrote to the psychologists Alfred Binet, Hans Gross, and Édouard Claparède to ask whether the techniques Münsterberg championed "have been used in a single instance in a European trial" or had been "sanctioned by a European Supreme Court." The answer was no.[48] He concluded that experimental psychology was not precise enough to do the job. Its vaunted ability to evaluate individuals' minds was in reality too slight to justify the reform of legal practice. And the adversarial system provided a more than sufficient incentive to litigators to use a tool like experimental psychology if they could truly make it work in a courtroom. He therefore found Münsterberg guilty of inflating the merits of his own field while libeling the law, and of creating conflict where he should be trying to encourage a "friendly and energetic alliance of psychology and law, in the noble cause of justice."[49]

From Warnings to Promises

Münsterberg's failure to establish forensic psychology in the courtroom, despite his popular appeal and the (qualified) interest of legal heavyweights like Wigmore, stemmed in part from a more general predicament facing would-be psychological experts. They had to win over a public that saw unbounded potential in the powers of science but resisted expertise when it threatened the sovereignty of personal judgment.

In the early twentieth century there was an enormous optimism about the "magical" powers of science to solve any problem.[50] Although the new psychological sciences were hugely popular, one result was confusion between "legitimate" and "illegitimate" scientific authority, accentuated because scientific and medical writers of this period often ventured far beyond their educational or professional specialties. The circumstances that made possible Münsterberg's success as a popular writer also undermined his attempts to get people to defer to his own expertise.

The courtroom was a dangerous place to try to establish one's expertise. The public standing of nineteenth-century doctors had suffered when they ventured into forensic testimony.[51] The challenge facing forensic psychologists was all the greater because their expertise did not relate to things that juries considered themselves incompetent to understand—the composition of blood, for instance, or the cause of death. Münsterberg and other psychological experts asserted that juries were incompetent to perform their basic function—evaluating what witnesses said using ordinary common sense—because common sense alone was not up to the job. Thus he raised the hackles of lawyers—and the general public—who were not about to cede the authority of commonsense judgment to this extent.

Yet the careers of memory techniques like Münsterberg's cannot be explained merely by recourse to a populist rejection of elite intellectuals or the self-interest of professionals. Most of Münsterberg's claims were very popular with non-academics—at least with his actual readers, the middle-brow subscribers to *McClure's* and the *Atlantic*, whose editors consistently begged Münsterberg for more articles. The kinds of topics he wrote about before his memory project were far more intellectual and even abstruse—art education, idealism, and American "traits" as viewed from the perspective of a German observer.[52] Memory was in many ways more accessible. All readers had memories and could reflect on their own experience, and many did so sympathetically. It is also worth noting that although Münsterberg wanted most of all to reach lawyers, he did not try to reach them directly. He asked lay readers to demand that the courts make use of experts like him. The success of Klein's play suggested that at least some of Münsterberg's claims took hold. He himself reckoned that his success with readers on other applied psychological issues augured well for this strategy, and reviews in popular magazines certainly were better than legal reviews.

The problem with Münsterberg's claims was most likely not that they came from academia. His previous writings had leveraged the laboratory to educate general readers. But this work had not addressed a subject on which each person considered himself an authority. Everyone had at least a nascent opinion about memory, and in *this* setting they all had powerful reasons to hold to what they knew. Implicit understandings of memory validated people's confidence in what they could know and even who they were. Münsterberg's own burglary story conveyed an unsettling process of self-discovery: he was not the person he thought he was. He knew less about his personal past, and even how much he *could* know about his personal past, than he had assumed. For Münsterberg, this was a preamble to a claim to greater knowledge. Through the failure of his own memory he came to know more about memory in general. The lesson for him was a positive one, but it was not so encouraging for his readers. They were told to doubt what they could know about their own pasts. But doing this did not make them more competent in their memories. Instead, they were to consult someone like

standing psychiatric practices, and they had a cultural traction that has never really disappeared.

From Twilight Sleep to Truth Technique

Robert House was among thousands of obstetricians who tried "twilight sleep" in the mid-1910s. House, an obstetrician based in Ferris, Texas—a little farming and brick-making town near Dallas where he had settled after taking his MD from Tulane in 1899—used a practice commonly called the "memory test" to calibrate the dosage. He showed the patient an object before receiving the drug and then again after each of successive doses; the correct dose had been reached when she could no longer identify the object. Scopolamine's anesthetic efficacy was therefore calibrated through its *erasure* of memory. But House eventually came to believe that the drug could also achieve what might seem like the opposite result. It could make subjects relate the content of memories that were very much intact.

According to House's own account, the memory test did establish that patients attained a deep state of unconsciousness. He noticed that they rarely remembered the labor. Then, however, came an occasion in 1916 when something strange occurred. He asked a patient's husband for the scales to weigh the baby and was told they could not be found. The patient suddenly spoke, giving precise instructions for finding them. House was struck because the woman was deeply unconscious yet could answer questions. He decided to make further tests and concluded from them that scopolamine could be used to extract truthful statements—even from people who might otherwise lie. He became convinced that with this drug he could "make anyone tell the truth on any question."[7] Twilight sleep could therefore be put to work as a forensic tool.

House's story raises a great many questions. The most immediate is how he came to make any forensic associations at all. The woman's speech was in no obvious sense a confession. What impressed him was her ability to communicate at all—because she had no incentive to lie. One source of inspiration might be an old convention from the colonial period, when midwives asked laboring women about the paternity of their children, assuming that this was a time when they would not lie.[8] I have found no evidence, however, that twentieth-century doctors even knew of this convention. Another possibility is that House, who is still remembered in his hometown as the "perfect gentleman,"[9] heard something he did not feel he could repeat in print.

What is certain is that the effects of scopolamine were being closely watched in 1915–16 because of concerns about its safety. The drug's popularity had waned soon after its introduction in the early 1900s, with reports of complications and the deaths of several babies. It then enjoyed a dramatic resurgence in 1915 after an article in *McClure's Magazine* promised it was safe when properly administered. Mothers-to-be then demanded a ride on the "sleeping car 'twilight.'" In late 1915 and 1916, doctors were giving it to their patients but nervously monitoring the effects.[10] House may therefore have had reason to pay special attention to the nuances of his patients' behavior under its influence.

These years also saw a surge of publicity for lie-detection technologies, which could have primed House to make specifically forensic connections.[11] If this was so, however, why did he choose to announce scopolamine's forensic potential only in 1922, some five years later? House himself gave no explanation, but the reason perhaps lay in public concern at that time about an apparent upsurge in both official corruption and crime. The late 1910s and 1920s were marked by commitments to deal with dishonesty: juvenile delinquency, organized crime, and even apparently endemic corruption in the police and government communities themselves. Editorials called on political reformers and scientific innovators to address the "endless chain" of graft that was "stalking the land."[12] In the months before House's announcement, Dallas papers were calling citizens' attention to a wave of crime so intense that it threatened the very social order.[13] Police were urged to become more scientific, and everyone was told to worry about the validity of confessions and eyewitness identifications.[14] Every day, therefore, House's newspapers told him about efforts to make police work more scientific, but they also told of a tenacious culture of crime that seemed only to get worse. In early 1922, editorials urged readers to address the challenge of crime. Citizens were to ask themselves, "What have I done to help?" They should visit prisons to see conditions, participate in meetings to support Prohibition, and encourage other reforms that would increase public order and probity.[15] House effectively answered that call.

The first public trial of scopolamine was carried out against this background on February 13, 1922, on two prisoners in a local jail.[16] One was W. S. Scrivenor, a confessed member of a gang that robbed a post office in Dallas; he wished to refute claims that he had carried out another robbery. The other was Ed Smith, a "negro" accused of murder, who was protesting his innocence. News reports regarded Scrivenor's confession to the Dallas robbery—and the fact that the crime at issue would not have an effect on his

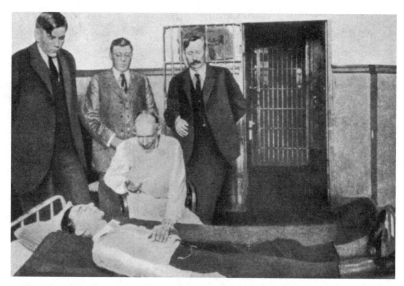

Robert House experimenting with scopolamine on a prisoner in a Dallas county jail (1922). From Emilio Mira y López, *Manual de Psicología Jurídica* (1932).

jail time—as evidence that his testimony about the other robbery was not self-serving. After the experiments, he described his experiences in a statement whose terms could have been supplied by House himself: "Answers to questions slipped from my mind without any apparent desire to stop them, and I felt that I couldn't formulate any imaginative trimmings to them."[17] He could do nothing, he said, but answer the questions truthfully, "knowing that I was telling the truth, and that it was impossible to do otherwise."[18]

On the face of it, Scrivenor's statements were extraordinarily well suited to House's goals, but they also carried a potential danger. Scrivenor claimed he had been compelled to answer "truthfully." Yet his saying that he "knew" he was incapable of lying, along with his reflections on his memories, indicated a degree of self-consciousness that potentially undermined House's claim about how truth serum worked.

What could explain scopolamine's power? After it began to be used in a forensic context, researchers paid more attention to obstetric patients' behavior. It then emerged that not only did they lose their memories *after* the treatment, they also experienced a lightening of the burden of pain of all kinds—emotional as well as physical—*during* the session. Patients might remark that their pain (or sadness, or fear) was intense, but they seemed not to be bothered by the pain itself. It seemed to witnesses as if the knowledge of the presence of pain was severed from the experience of it, and the experience itself was somehow suspended. This seemed to explain why patients would part with damaging or upsetting information in forensic exams: they could not "feel" the painful implications of doing so. This would mean that scopolamine did indeed act as a kind of anesthetic and not just as a memory blocker. No one remarked on the subtle and anti-intuitive features that made it effective as an anesthetic until the forensic part of scopolamine's career. Ironically, then, a central feature of what was thought to make scopolamine effective as an anesthetic contributed to the (superficially contradictory) notion that it could extract memories.[19]

House's claim, based partly on this reasoning, was that the conscious mind was disabled, or perhaps bypassed, by the drug. Yet Scrivenor's account of his experience seemed to contradict this. In fact, however, neither House nor anyone else commented on this feature of Scrivenor's case. The reason could be simple self-interest on House's part. But House was not a scientific researcher or an academic; he was not interested in psychological or physiological questions for their own sake. He wanted a tool to address intractable problems of crime management that had come to his attention.[20] At any rate, this lack of curiosity seems to confirm that his concern had shifted decisively from scopolamine's anesthetic properties (and hence its amnesiac influence) to its newly identified power over memories.

Scrivenor's testimony helped to establish scopolamine's plausibility as a truth technique, since as a confessed bank robber he seemed to confirm that a criminal would indeed make a true confession when conversing under its influence. This was significant, since Ed Smith, who persistently asserted his innocence both before and during the scopolamine test, thereby gained in plausibility. Smith had been arrested in 1921 for a 1916 murder, the primary evidence against him coming from witnesses who turned state's evidence. His protests became all the more impressive when these witnesses refused to take the scopolamine test. He was eventually released.[21]

Truth serum immediately began to draw national attention. It became a fashionable topic of discussion in medical, psychiatric, and legal societies in Texas, in the Midwest, and soon on the eastern seaboard.[22] House himself published his initial findings shortly after these experiments, and he followed this first paper with a series of others that inspired national controversy. Legal journals and popular magazines variously hailed his work or decried it as suggestive or unpromising, "revolutionary" or ridiculous.[23] The term "truth serum" was coined within the first few months, supposedly by a Los Angeles newspaper,[24] and it subsequently proved irresistible even to doctors who warned that scopolamine was neither a serum nor a straightforward extractor of "truth." Meanwhile, House entered into "communications" with "medical authorities in Chicago," home of leading forensics experts and from the point of view of Dallas a great metropolis. He was about to step from the backwater into the mainstream.[25]

Truth Serum on Trial

The early 1920s were a significant moment for "truth" techniques in general. The Frye ruling of 1923, which set the standard for the admissibility of scientific techniques for most of the twentieth century, was a judgment in a lie-detector case. In the years following the initial publicity over House's technique, it was consistently snubbed by the courts as insufficiently proven to be admissible as evidence. But at the same time, truth serum was welcomed into the court of public opinion, debated in the legal and forensic press, and increasingly used in informal public and juridical contexts. Police, prison officials, chaplains, and some (though few) lawyers—but no judges—found the notion of a truth serum plausible and attractive in principle, and they linked their own conventions for evaluating truthfulness to the action of the new chemical technique. Truth serum also seemed to accord with commonly held understandings of memory, volition, and truth telling. The antipathy of elite expert communities (such as academic psychologists and senior jurists) had little impact on the popular enthusiasm for it.

One of the first and most spectacular cases of such use involved a string of two dozen ax murders in Alabama. Several people submitted to scopolamine interviews in 1924; five of them confessed under the influence of the drug and confirmed their confessions after waking from the trance.[26] What credibility this case gave to scopolamine was mitigated, however, by its subsequent use in a Hawaiian kidnapping and murder case. Here a chauffeur confessed during a scopolamine interview, then retracted his confession. A second scopolamine interview produced negative results. At this point police arrested another suspect who ultimately was convicted of the murder.[27]

House himself toured the country several times carrying out "scopolamine interviews" on criminal suspects and convicts to demonstrate the potential of his drug. Among his many stops were San Quentin Prison in California,[28] Los Angeles, and New Orleans, where city reporters described dramatic confessional effects on themselves when they submitted to the experiments, but more ambiguous results on prisoners in the local jail.[29] By 1925 House could claim he had made eighty-six trials of scopolamine on criminal suspects or convicts. Although this information was never admitted into evidence in court, in twenty-six of these cases the new information led to their release.[30]

One of House's stops was at a string of prisons in Missouri. The St. Louis newspapers provide an especially rich portrait of his experiments and of the discussions that followed them. First he visited the Missouri Penitentiary in Jefferson City to carry out a scopolamine interview on Martin Hulbert, a convict serving a life term for murder. Hulbert, who had requested the trial, had been convicted on circumstantial evidence, and his case had twice been appealed to the state supreme court.[31] Local newspapers quoted the interrogation:

"Who did you kill?"
"I didn't kill anyone."
"What kind of a gun did you shoot him with?"
"I didn't shoot anyone."
"Who arrested you?"
"Officers Smith and Malone."
"How long after the killing were you arrested?"
"Three days."
"Did you have a pistol when arrested?"
"I never had a pistol."

The test continued for an hour, pausing when Hulbert appeared to be exhausted. Local papers recorded that he "declared his innocence" before members of the state penal board, but also that the courts would not accept the experiment as the basis for an appeal.[32]

House was then invited to St. Louis by a prison chaplain, the Reverend John A. de Vilbias, where he conducted demonstrations on three prisoners and one journalist. Present were De Vilbias, officials of various St. Louis institutions, the circuit attorney, the prosecuting attorney, the circuit judge, and several psychiatrists and doctors. The lawyers expressed an interest in scopolamine, but the circuit attorney said the results would have no effect on any of their cases. Then the drug was administered and each man was tested using the "memory technique" to calibrate the dosage. All four, according to a local paper, acted "like drunken men." Then they were interviewed, with detailed questions about their alleged crimes. In this case the newspaper maintained a healthy, though respectful, skepticism. It had been "difficult to get [subjects'] attention or to make them understand, and it was obviously difficult for them to make coherent replies." House himself was unsatisfied with the experiment, though he attributed the problems to the distractions of the crowded room.[33] Some witnesses sensed that, while the drug seemed to break down "reserve" and make people "less likely to tell lies, it has also disabled them mentally to such an extent that they could not comprehend or utter very much of the truth."[34] Others, including De Vilbias, pronounced in favor of scopolamine.

At least one of these experiments led to an attempt to change the course of a trial. This was the 1926 case of George Hudson, a "negro" accused of raping "an aged white woman." The victim identified Hudson as her attacker. He pleaded mistaken identity, and witnesses swore that he was elsewhere at the time. The defense enlisted House, who carried out a scopolamine interview in preparation for the trial.[35] In a deposition, House described the physiological state created by scopolamine, the tests he carried out on Hudson, and his own consequent belief in Hudson's innocence. The transcript of the interview was presented by Hudson's lawyer as part of the defense. But the judge rejected it as scientifically unproven, under the authority of the recent Frye ruling on scientific evidence, and Hudson was convicted. In the appeal, two years later, Hudson's counsel argued that the deposition and scopolamine interview had been improperly excluded. This time the summation by Judge Robert Walker Franklin not only upheld the original decision but scathingly dismissed House's approach:

> Testimony of this character—barring the sufficient fact that it cannot be otherwise classified than as self-serving declaration—is, in the present state of human knowledge, unworthy of serious consideration. We are not told from what well this serum is drawn or in what alembic its truth compelling powers are distilled. Its origin is as nebulous as its effect is uncertain. A belief in its potency, if it has any existence, is confined to the modern Cagliostros, who still . . . cozen the credulous for a quid pro quo, by inducing them to believe in the magic powers of philters, potions, and cures by faith. The trial court, therefore, whether it assigned a reason for its action or not, ruled correctly in excluding this clap-trap from consideration of the jury.[36]

The high enthusiasm for scientific experts at this time carried its own risks—along with characters like Sherlock Holmes came a category of Cagliostros and Svengalis. House's challenge was to look more like Holmes than Cagliostro, and in this case he failed.[37]

Truth in Memory

What well *was* truth serum drawn from? In each case he attended, House gestured toward a physiology of memory. He drew on a common idea of that altered states of mind are characterized by psychological automatism—an incoming stimulus (the question) reflected back in a reply without the intervention of the will. There was a long tradition of the notion that alcohol loosened an individual's self-control (one marker of it being Pliny's famous dictum). Victorian psychophysiologists developed a more specific notion of automatic mental response during an effort to explain quasi-hypnotic states. By the end of the nineteenth century, it was generally agreed that in certain psychic states it was possible to process questions "automatically," as a reflex.[38] In this state, House wrote, "the stimulus of a question can only go to the hearing cells." Then "the answer is automatically sent back, because the power of reason is inhibited more than the power of hearing." He called this state "House's receptive stage."[39]

House provided a more expansive account in a deposition for the Hudson case. When a question is asked, he explained,

> the sound waves hit the drum of the ear, and the nerve of hearing, like a telephone wire carries the sound waves to the center of hearing. The only function of the center of hearing is to evoke memory by sending the impulse to that part of the brain where the answer is stored for future use, and . . . the brain sends the answer to the nerve of the tongue. To illustrate—If I ask a person "What is your name?" he cannot by the will power or by any other function of the brain keep from . . . thinking the answer, but his will power can prevent the tongue from articulating the name, and the power of reason can also take the answer and be calling on the imagination make the tongue tell a lie, but when the will power and the power to reason are removed, the replies are automatic. Hence the truth. When the will and the power to reason are nonexistent, then man is too unconscious and too helpless to prevent himself by inventing replies to questions propounded.[40]

He never addressed one aspect of the scenario that might seem contradictory: How could a drug known for its power to suspend mental functioning force the mind to communicate reliable memories? House himself used the "memory test" on truth-serum subjects as well as obstetric patients. He must have expected to produce memory loss for events in a subject's *immediate* past. This suggests an implicit distinction between very recent and longer-term memories, such that scopolamine could demonstrate its efficacy in the memory test yet still faithfully deliver intact long-term recollections. House was not relying on a formal distinction between short- and long-term memory—this did not yet exist—but rather posited an intuitive one that he never explicitly articulated.

Modern readers may wonder whether, under truth serum, one's imagination could produce artificial memories, with or without inadvertent suggestions from the questioner. Despite Hugo Münsterberg's warnings back in 1909, this issue was not as prominent as one might expect. House was concerned neither about suggestion nor about the possibility that memory might be anything more artifactual than a permanent trace of original experience. This trace was the "truth" that House held out to the public as being accessible via scopolamine. He did consider the possibility that the mind might "invent" memories, but he did so only in passing and only in order to deny that a drugged mind could do such a thing. Drugged individuals were merely conduits for information, with "no power to think or reason," because "the conscious mind is asleep while the subconscious mind will function by stimulation of the center of hearing." In short, "under the influence of the drug, there is no imagination."[41]

House himself gave no more sophisticated or explicit account of "automatic" mental actions or memory than this. This may have been to his advantage: his vague claims chimed with the statements of other researchers. It also seems to have drawn on popular assumptions about the will and memory, since the press treated his claims as intuitively plausible. What was amazing, these articles suggested, was the development of the truth-serum technique, not the existence of reliable, truthful memories as entities within the body that were in principle accessible to such a technique. The *St. Louis Post-Dispatch*, for instance, explained that the "hearing was unaffected by the drug," but mental functions were so disabled that the will could not "distort facts imprinted upon the brain of the subject."[42] Forensic professionals soon took up House's claim, echoing it in roughly the same terms. For example, the highly influential forensic pioneer Frederick Inbau wrote that "various drugs are capable of producing a mental state in which consciousness is more or less profoundly affected, thereby rendering a suspect's reactions somewhat automatic."[43]

Implicit in truth serum was an assumption that each event, experience, or thought left a corresponding trace that could be called to consciousness when needed. The original record did not change, except to become weaker. This was a popular view of memory at the time when House was developing the concept of a truth serum, but it ran directly counter to the work of the community that saw itself as most knowledgeable about mind and memory—the psychologists. Since the early days of laboratory psychology, academic psychologists had found myriad ways of documenting the changeable character of memory. In 1895 psychologist J. M. Cattell documented errors in particular types of observations and recollections, constructing a sort of taxonomy of witness fallibility; and long after Münsterberg's work of 1906–8, psychologists regularly studied the fallibility of memory.[44] In 1920 in Britain, Sir Henry Head discarded the notion of traces altogether and argued instead that every experience, understood as a "postural change," altered the personal set of references that we used to understand our relationship with our immediate past and the world around us. New changes were always "measured" against those that came before: "By means of perpetual alterations in position we are always building up a postural model of ourselves which constantly changes."[45] By the mid-twenties, psychologists could point to decades of experimental evidence showing that a truth serum should be impossible in principle.

The explanations associated with truth serum therefore ran increasingly counter to the grain of academic psychology, whether in America, where it was soon dominated by behaviorism (whose proponents were opposed to thinking about memory in terms of ideas and meaning rather than behavior), or in England, where a highly constructivist—and eventually hugely influential—account of memory was being developed by Frederic Bartlett in Cambridge. This accentuated an existing divergence between public and academic representations of memory and psychological truth.

A "Harmless Third Degree"

Truth serum eventually became identified with involuntary confessions wrung from unwilling subjects. But this was not its initial purpose. House himself regarded the serum as an important tool for compelling honesty in *institutions* as much as in individuals. The vindication it could offer to the falsely accused would, he thought, force transparency on a corrupt criminal justice system poisoned by a culture of graft and private deals. House offered

A criminal's craving for drugs provides an opportunity for a truth-serum interrogation. *Dick Tracy* comic strip, *Chicago Tribune*, September 19, 1933. © Tribune Media Services, Inc. All Rights Reserved. Reprinted with permission.

different emphasis. The idea that in certain circumstances recalling could be a simple reflex arc relied on the notion of the will as a censoring influence. It was the existence and power of the will in one's ordinary state of mind that allowed for reflexive communication in circumstances when the volition could be suspended. In House's work, the suspension of the will was a means of confirming that one's will was aligned with one's self-representation. In the newer, more combative formulation of truth-serum techniques, the assumption was that the will was an oppositional power that had to be removed to expose a concealed truth.

Truth serum now became a means of consolidating the powers of an authoritative police and judiciary rather than a tool for reforming them. Celebrations of this power to "get the confession" soon brought anxious reflections about the implications for civil liberties and personal privacy. It was in the 1930s, then, that "truth serum" acquired its association with abusive interrogations and involuntary confessions.

An important factor in this transformation was the work of the Scientific Crime Detection Laboratory (SCDL) at Northwestern University. Founded in 1929 in the wake of the St. Valentine's Day Massacre, the Scientific Crime Detection Laboratory stood at the vanguard of a broader effort by specialist laboratories and forensic techniques to introduce medical expert witnessing into the criminal justice system.[53] The new forensic laboratories of the 1910s and 1920s made the first laboratory tests of the polygraph and of hypnosis; developed techniques for chemical, fiber, and ballistic analysis; and found new ways of evaluating fingerprints. Their staffs formed a new cohort of expert witnesses. Northwestern's laboratory was to be a flagship institution in this forensic science. It viewed forensic tools and techniques as extensions of legitimate, expert authority, arrayed against a target body of straightforwardly criminal "elements."

Landmark legal rulings came about partly as a result of these new forensic institutions and techniques. The most famous of them included the Frye ruling, excluding techniques that had not yet attained general scientific acceptance, and the "exclusionary rule" disallowing the use of illegally obtained evidence in criminal trials. In less than two decades these rulings fundamentally changed the relation between legal and scientific expertise.

In this context, what counted as a scientific technique tended to vary. Some forensic scientists, like House, represented their work as an alternative to the third degree—a scientific, intelligent procedure instead of a crude practice based on bullying or even torture. But even the third degree could be portrayed as scientific. Some news articles on forensics represented it as a collection of precise psychological techniques designed to weaken the nervous system so greatly that it could not withhold information. One author wrote that in "popular" language it was a scientific way of "getting on one's nerves."[54] On this account, truth serum claimed to achieve by the immediate action of a chemical preparation what the third degree sought to secure using verbal, social, and physical pressure: an alteration in nervous functioning, leading to an honest statement. Fred Inbau, a leading figure in the Northwestern lab, thus aligned scopolamine interviews with police interrogations: "The policeman with his 'third degree' and the scientific investigator with his 'truth serum' are both working toward a common objective."[55] And the work carried out by the SCDL was critical in enhancing forensic drugs' reputation as producers of truthful information. Key researchers here were the chemist Clarence Muehlberger (destined to have a long and prestigious career in forensics), the ballistics expert Calvin Goddard (then head of the lab), and the lie-detector expert Leonard Keeler.

One of the early experiments at the laboratory lent particular support to scopolamine. The test group of students and staff made up lists of questions based on their own pasts, complete with true answers. The subjects attempted to lie when answering these questions under the influence of the drug. The results, as they were recorded in published accounts of the case, gave the drug nearly a perfect score: every answer on every list matched the original answers. There was a single exception—the answer to just one question had been wrong. Such a small degree of error was, of course, striking, but the experiment turned out to have been even more successful than was initially appreciated, because the solitary mismatched answer turned out in fact to be a true statement. The question had been, "Were you ever arrested for a motor vehicle infraction?" The subject recorded no initially, then yes during the scopolamine interview. After he recovered from the effects of the drug, he was asked about the discrepancy. He reportedly answered with some embarrassment that yes, his drugged answer was correct. When he composed his list of questions he had forgotten that many years earlier he had indeed been arrested while driving. The story supplied an elegant piece of support for scopolamine's advocates.[56]

Northwestern also used scopolamine in its investigation of criminal cases. One of its biggest early successes involved a man accused of murdering his lover's husband in 1934. The police suspected the wife and her "paramour." He agreed to take a scopolamine test, and while under its influence, he was asked what he "did with his pistol." He replied that he "threw it into a river." This was a confusing reply, since the victim had been found dead in bed with the gun beside him. When the question was repeated, he elaborated, saying he "hid the gun in a patch of heather in a town in Ontario, Canada. Concerning the present crime, however, he continued to profess his innocence." The police checked with officials in Canada and learned that he was wanted there for several murders.[57]

This and several other high-profile cases gave truth serum a controversial, but highly visible, role as an aid to interrogation and associated it with an armory of powerful forensic tools. At the 1933 Century of Progress Exposition, the five stalls of its "police exhibit" included a "physiological laboratory" stocked with lie-detector devices and equipped for a staged series of tests, including demonstrations of truth serum by Muehlberger.[58]

The Northwestern laboratory was intent not just on boosting the reputation of its new tools, but on training the police and legal communities to respect and use them. Its training manual, as well as syllabi for lectures given by Keeler, Inbau, and others, portrayed truth serum as an effective tool for securing confessions.[59] The SCDL was extremely energetic in these training and demonstration programs, and through its activities a new generation of law-enforcement professionals was given a small dose of training in the use of truth serum along with its more general forensic training. There are accordingly signs that police forces used truth drugs regularly during the 1930s. The Wickersham Commission's 1931 report on police behavior stated that criminal suspects received scopolamine (along with "tear gas" and chloroform) during interrogation. One policeman noted that the mere threat of such a test could secure confessions:

> One department demonstrated to suspects just how the drug was administered, and went on to explain just how it worked. . . . If the suspect said that they couldn't force him to submit to the drug, it was pointed out that the drug also worked when the suspect was first rendered unconscious by chloral hydrate . . . placed in his coffee or drinking water. In many instances, the terrified suspect talked to avoid being tested with the truth serum.[60]

The Northwestern laboratory's effort to advance truth serum as a scientific way to "get the confession" was clearly effective. No matter how "sternly or skillfully [the culprit] represses every outward sign of guilt," wrote journalist Henry Morton Robinson in the *Forum*, "he cannot prevent the involuntary mustering of interior forces." Robinson unequivocally described scopolamine as "a mysterious elixir that gets the whole story and gets it straight." It worked, he argued, by "submerging" that part of the brain that was "normally used in fabricating self-protective stories—in plain language, lies." And so "a person under the influence of scopolamine is in full possession of all his senses, but deprived of the power of inventing falsehoods."[61] This account portrayed science as a delicate yet powerful instrument, replacing the blunt force of the cudgel or rubber hose. Yet powerful opposition to the technique remained. Just as often, medical and legal essays portrayed truth drugs either as inadequate to the forensic task or as a thinly disguised form of torture. House's dream of a pharmaceutical cure for the criminal justice system began to fade as "truth serum" began to be more commonly associated with dubious ways of pressuring witnesses into disclosing their memories.

The Transformation of Falsehood

By the 1930s, truth serum had become fairly established, and not only in the public imagination. It was beginning to be taught in leading forensics laboratories, even though it had been soundly rebuffed by the courts. Its ascent during these years has something to tell us about the categories of "expert" and "popular" knowledge in a period of rapid professionalization and formation of new disciplines in all the communities involved in this history—psychological, medical, forensic, and police. The assumption promoted by the laboratories and departments of psychology that were built in the 1900s to 1920s was that trustworthy knowledge of human behavior could be gained only by individuals trained to ask questions and make disciplined observations in the ways sanctioned there. Psychologists often said that the goal of laboratory psychology was to "objectify" the human mind. The drive for professionalization involved campaigns against illegitimate practices and practitioners, notably the many waged by the American Medical Association against "quacks" and "quackery"—not least against twilight sleep itself. But scopolamine's associations were multiple and conflicted. House himself had never seen a psychological laboratory, and his claims about scopolamine were not based in his area of medical expertise. That he could enjoy such a welcoming reception suggests a certain malleability in how one could plausibly announce new discoveries, in exactly the period when the AMA and other professionalizing institutions were pressing home their campaigns.

Yet this does not mean that just anyone could come out of nowhere—as House virtually did—and be taken seriously. House was an educated man and a member of the professional classes, and his business was the human body. He had enough claim to legitimacy that he could hope to parlay his expert qualifications into a successful venture in this new field. And this was not such an unusual tactic at the time. Inventing a new technology, technique, or approach to a recognized problem seems to have been a common way for an educated person to establish a claim to special knowledge.

Truth serum's later history was rather peculiar, seeing neither increasing legitimacy nor consistent decline. On the contrary, its appeal spiked at various times of urgency, as we shall see in later chapters: in World War II, when truth drugs became central to the military treatment of battle trauma; in the early 1960s and mid-1970s, when forensic psychiatrists and police seized on forensic hypnosis and hypnotic drugs as tools for refreshing witnesses' memories; and in the 1980s and 1990s, when psychotherapists and their adult patients turned to the drug to help in the "recovery" of lost memories of childhood sexual abuse. At other times it languished with little respect or credibility. Truth serum's liminal existence allowed it to wax and wane with the desire that successive generations and distinct communities felt for particular kinds of personal memory.

3

MEMORIES OF WAR

During World War II, soldiers throughout Europe struggled to recover from terrible battlefield traumas. In many cases their worst injuries seemed to be not wounds or broken bones but experiences so terrible that their minds could not accept them as ordinary memories. While this information remained unprocessed, it was all the more potent as a result, causing powerful psychiatric and physical symptoms. Military psychiatrists tried to address those symptoms by bringing the past back to life.

To treat these troops, doctors used barbiturates and other ways of inducing altered states of mind. Under their influence, a patient would reexperience the sights, sounds, smells, and emotions of a particular moment in the past under the watchful eyes of a psychiatrist. This cathartic practice spread through the army, where it was celebrated as the linchpin of humane yet speedy psychiatric care. The idea of accessing blocked memories motivated the work of hundreds of military psychiatrists and field doctors and the experiences of thousands of patients. It was also central to widely disseminated training films, documentaries, and dramas.

The "truth serum" drugs of the 1930s, sodium amytal and Sodium Pentothal, thus became the "narcoanalysis" drugs of the 1940s. Hypnosis was used for similar purposes. The people who used these techniques at midcentury were not criminal justice professionals but psychiatrists; their goal was not to get at legally significant information but to heal troubled minds. But although the assumptions underlying narcoanalysis were a departure from those for truth serum, narcoanalysts and forensics experts agreed that somewhere in the mind lay an intact record of past experience, if one could only get at it. Robert House had likened memories to papers in a drawer, sliding open in response to the right pharmaceutical key. Midcentury psychiatrists were more likely to compare memories to motion picture films. Sometimes the rememberer was a member of the cast; at other times he was in the audience. Either way, memory now belonged more to the world of film information schemes like Vannevar Bush's Memex than to that of earlier technologies like the filing cabinet.

Abreaction

The idea that long-forgotten traumatic events can suddenly come back to life did not originate in World War II. During and after World War I, psychiatrists treated many soldiers suffering from amnesia and other apparently psychosomatic problems by using conversation, hypnosis, and drugs to excite a sudden return of memory. The hope was that this "abreaction," as psychoanalysts called it, would make possible the beginnings of a recovery. Psychotherapy would help patients integrate the newly remembered experiences into healthy memories.[1]

During the 1920s and 1930s psychiatrists used the new barbiturates on nonmilitary patients, hoping to reveal "hidden sources of conflict" buried in these patients' minds.[2] In 1938, for example, psychiatrist Morris Herman gave sodium amytal to a young woman found wandering the streets in a "sort of stupor." She immediately recalled her name and address. Then she disclosed intense "guilt feelings for an avowed Lesbianism," to which Herman attributed her amnesia.[3] Experiments of this kind became common, and soon they even had a name, "narco-analysis," coined by British psychiatrist John Stephen Horsley. Barbiturates uncovered the causes of psychiatric symptoms, Horsley believed, and then helped "rearrange" those causes. Soon after administering Amytal, the physician was to engage the patient's attention and establish a rapport. The patient would then "volunteer the information that he 'feels confidential,'" and the physician could begin analysis. Through free association or other techniques, memories would be "synthesized" into consciousness. Once the "salient factors" responsible for the patient's condition were first identified, the patient could forge new associations to the traumatic memories. Patients did not always need lengthy treatment: Horsley sometimes achieved cures "at a single sitting."[4]

Horsley gave a number of examples, beginning with a case of a soldier who had suffered from war trauma for fifteen years. His worst symptoms were convulsions, following a terrible experience of which he had no memory. Under narcosis, he "recalled a vivid picture of his war-time experiences up to the time of a bayonet attack." He described terror, then loss of consciousness. Horsley told him that he would "become convulsed" one last time, then awaken cured, with his memory restored. He did indeed become convulsed, then shouted as he "seized a phantom rifle and fought for his life against invisible foes." After another dose, he slept all night and woke with complete recall.

Amytal worked, Horsley wrote, because it removed the inhibition of repression, restoring patients' access to painful information. He did not consider the possibility that his techniques might lower the "threshold of recall" for events that were fading from the mind in a natural process of attrition. This would have assumed that the mind discards many memories and that their loss does not necessarily serve some emotional purpose. Like analytically oriented psychiatrists, he assumed that experiences were retained in memory indefinitely. Although his work did not attract much notice initially, it caught the eye of military psychiatrists planning for large numbers of casualties. By the early 1940s it was already becoming a classic of modern psychiatry.

Sedation and Adjudication in Surrey

At the beginning of the war, British military leaders expected a flood of psychiatric casualties. Several hospitals were equipped with hundreds of beds and staffed by experienced psychiatrists. Casualties did not arrive immediately, but after the evacuation of Dunkirk in 1940 they poured in. Many of them came to William Sargant and Eliot Slater at the Sutton Emergency Hospital in Surrey. They treated patients with a variety of techniques, including insulin coma, electroconvulsive therapy, and frontal leucotomy. Sargant also used psychoactive drugs—especially sodium amytal—to deliver long periods of rest in "battle exhaustion" cases.[5]

Sargant was no psychoanalyst. He rejected psychoanalysis wholesale, believing that psychiatry should address itself to the "physiological components" of mental attributes like "courage, will-power, and self-control." Psychological problems were for him the result of an altered *physiological* state, even if they produced *mental* abnormalities. Drugs were useful not to facilitate a certain kind of thinking, but as sedatives. While psychoanalysts de-

Sutton Emergency Hospital after bomb damage, World War II. Courtesy of Ben Wood.

scribed drugs as "psychically 'analgesic'"—lessening the pain of emotions[6]—
Sargant's references were literal. He liked to say that "one feels afraid with
one's belly and not with one's brain." When the body was strengthened, the
mind would follow.[7]

Yet although Sargant claimed not to credit the power of talk, his case notes
tell a different story. He carefully quoted patients' words under Amytal and
wrote that one could reconstruct lost "information" using hypnosis. Although
the result was "a mixture of truth and fantasy," one could "sift" through the
patients' statements for the truth. This could lead to an "emotional synthesis"
and to "normal physical good health."[8] Sargant even used psychoanalytic
terms like "abreaction" and referred to "unconscious" memories severed from
their emotional "affect." It is possible that Sargant's own experiences with
Amytal affected his attitude. When bombing became particularly intense
(the hospital itself was hit before the end of the war), psychiatrists themselves were as likely to be high on Amytal as their patients were.[9] In any
case, by 1942, as the stream of incoming casualties slowed and the prospect
of German invasion became more remote, his treatments began to resemble
Horsley's.

Sargant's shift from his initial rest-and-feed policy to a more psychological one is clear in his dramatic film of 1942, *The Treatment of War Neuroses*,
which advertised his technique and was used to train residents for many
years.[10] It began with a history and a tour of the hospital, then began a chronicle of patient treatments, with special attention to one amnesiac patient. He
was a survivor of the RMS *Lancastria*, which was sunk by the Luftwaffe on
June 17, 1940, during the evacuation of France, with the loss of approximately
four thousand lives. It was the worst single event involving loss of life in British maritime history. So this patient represented much more than himself.

The film first shows him early in his treatment, as he slips into the "narcoanalytic" state after an injection. We then see him speaking, though his own speech is not audible; a narrator explains, "under the drug, inhibition goes. He is being persuaded to tell the story of his experiences during the sinking." The patient becomes more animated, and the narrator notes "the increasingly lively gesturing as his memory clears." A pair of "before and after" shots highlight the transformation: we see the patient tense and twitching before the treatment, and then afterward, calm and smiling. Then we see him a week later, looking cheerful and engaged in conversation. His interlocutor, apparently a therapist, is off-screen, and his absence encourages the sense that who he is does not matter—the patient can now interact with everyman.[11]

Sargant kept Amytal therapy in his psychiatric arsenal, but he found that in chronic neuroses he could not produce the "true" and transformative reliving of the traumatic event. In 1944 he began to use ether, after hearing that it could produce an extreme state of excitement and catharsis.[12] Ether also made it easier for Sargant to shrug off Amytal's psychoanalytic associations. Ether treatment made it harder for patients to talk, because ether was delivered as a vapor using a mask rather than intravenously, and also because of the intensity of the excitement brought on by the drug. Sargant believed that abreaction worked not because it brought an understanding of a past traumatic event (through the return of a memory) but because the abreaction event itself broke down unhealthy conditioned reflexes. Ether allowed him to adopt a Pavlovian interpretation of trauma and of the appropriate psychiatric response.

Sargant was following—or rather, helping to define—contemporary norms for what one would now call "standard of care." He was unusually prominent but was also representative of a well-accepted and popular approach to battle exhaustion, as dozens of articles in medical journals attest.[13] There was a range of opinion about what the drugs did and how effective they were, but within this range the use of barbiturates to produce memories was very popular. Not every psychiatrist was convinced that it worked; a Dr. J. L. Clegg complained that Amytal delivered nothing more than "ordinary conversation," and that at most it loosened subjects' tongues about memories of which they were aware but "ashamed." But such skeptics were soon in the minority. Clegg himself was roundly told off by a number of psychiatrists, among them Horsley himself, for being too inexperienced to carry out the procedure effectively.[14]

The memory therapy practices Sargant used in the early 1940s stayed with him in future years, when he continued to emphasize ether over Amytal. On occasion he was called in as a consultant to help revive blocked memories. For instance, while at a psychology conference in Boston in 1965, Sargant was approached to assist in the Boston Strangler serial killer case (discussed in chapter 6) by helping a witness identify her attacker. He used his ether technique, followed by a moment-by-moment staged reliving—to the point of placing his hands around her neck. It didn't work.[15] But for the most part his interests after the war lay in an altogether different area. Sargant was fascinated by the wholesale destruction of autobiographical memory, in the service of remaking an individual's personality. He became part of a network of psychiatric adventurers who believed that destroying memories, and even individuals' sense of who they were, could furnish a kind of tabula rasa on which to develop a healthy new self.[16] The extent to which Sargant himself actually executed such a program—commonly called "brainwashing"—is a matter of dispute. But it is clear that his interest in these extreme and controversial practices grew from the seed of his experience of psychiatric drug treatments in World War II.

Roy Grinker and Narcosynthesis

The American government did not follow Britain in anticipating significant psychiatric casualties when it entered the war. Only people with some underlying weakness would break down, it was assumed, and psychological screening tests would weed them out.[17] Since policymakers assumed that the few soldiers suffering psychiatric problems would be discharged, they did not plan for dedicated psychiatric hospitals, and few psychiatrists were assigned to the forward areas.[18] It soon became clear that these provisions were desperately inadequate.

Military documents divide the war into the mobilization years of 1940–42, when there was little psychiatric provision, and the combat period, 1943–45, when psychiatric services were established near combat areas and two large stateside psychiatric hospitals were founded. The policy change was a response to a series of increasingly panicked observations in 1943 that the armed forces were hemorrhaging psychiatric casualties, on course to lose half their manpower.[19] The discharge policy was frozen, and psychiatrists were exhorted to "salvage" every possible patient. With battles in Europe and North Africa producing thousands of psychiatric casualties, but with almost no psychiatrists there to treat them, the resulting scramble led to a disorganized, understaffed patchwork of therapists.[20] Into this chaos came a stream of doctors with neither military nor psychiatric background, who were pressed into psychiatric work.[21]

Roy Grinker was one of the few psychiatrists who enlisted early. After medical school in the 1920s and a psychiatric residency, he took a position at the University of Chicago, chairing the psychiatry department until the 1930s, when he became interested in psychoanalysis. He traveled to Vienna to undertake an analysis with Freud himself. When he returned, he left the University of Chicago (an inhospitable place for an analyst) to help found a program at Michael Reese Hospital a few miles away. When America joined the war, Grinker enlisted and asked to be sent to Britain.

But Grinker later recounted that he was put on the wrong ship, and some days into the journey he realized he was destined for Tunisia. He decided to make the best of it and was cheered to find that a colleague from Chicago, John Spiegel, had also been posted there. They set up a makeshift psychiatric hospital, and as they scrounged for medical supplies, they came upon a huge stash of Sodium Pentothal, then a popular intravenous anesthetic agent. Grinker was probably familiar with the work of Horsley and Sargant. He began to experiment with Pentothal at doses below the threshold for anesthesia.

Grinker had barely set up shop when the Tunisian campaign got under way. An intense series of battles began between Allied and Axis armies in North Africa. The fighting was particularly terrible for United States forces in March and April of 1943. A flood of traumatized patients swamped Grinker's makeshift hospital. Most were pilots, and fighter pilots in particular. This was a relevant factor in the kinds of mental illness they suffered, because pilots operated within a distinctive culture with its own discipline and social pressures. Fighter pilots were alone in the aircraft, so, according to descriptions of the experience, much of their discipline came from "within." There was a great deal of disciplining from without, too: great moral pressure was brought bear on pilots to complete their tours of duty. Anyone who did not want to complete his tour was stripped of his rank and branded a coward for the rest of his life.[22] This gave their psychiatric conditions unusual force. Grinker, facing an overwhelming influx of such cases, decided to give them Pentothal. He found that he was able to get some back to their units within a few days—a critical consideration in the circumstances.[23]

When Grinker began work in Tunisia, it must have been an extraordinary change for someone who had so recently spent months on Freud's couch and who was trained to think of therapy as a protracted series of expertly managed conversations. His new technique required almost no training or time. And the stories patients related under its influence were not the usual stuff of psychoanalysis—the childhood memories and intrafamily conflicts rendered symbolically in interesting symptoms. Rather, they were straight-forward narratives of events in the recent past. Therapists using Pentothal rarely dwelt on these narratives to probe their symbolic significance as a classical analyst might. Instead, they treated them as straightforward historical records. Classical psychoanalysts then (and now) might have looked askance at this project, criticizing it for naïveté both in the therapeutic potential of such speedy treatment and in the assumptions about memory that underpinned it, but Grinker claimed to be able to get some pilots back to the front in less than a week, as long as he saw them very soon after they became ill.[24]

The drugs' effectiveness reinforced for Grinker a simple, mechanical account of memory. He wrote that Pentothal caused the patient "to re-experience . . . traumatic battle experiences . . . which have been perpetuated in various stages of repression up to the moment of treatment." This enabled him to "deal with these revived emotions" in a "more economical and rational manner" than by earlier "catastrophic defensive technics."[25] Patients lay in a darkened room, on a slow IV drip, and soon became confused. Most then began speaking spontaneously; those who did not were to be prodded with a question or an instruction. The patient might be told that he was in the thick of battle: perhaps in flight, approaching the target, or with "flak bursting all about." Some psychiatrists intensified the realism by acting a part in the story, pretending to be a crewmate, "warning of an imminent ditching or asking for help with a wounded buddy."

The results were usually dramatic: some men so relived their experiences that they had to pace through them: "They may wander about the room as if about the plane, or, using the pillow or bedclothes as armor plate or some other protection, may wince and cower at flak and cannon bursts." Others just narrated the events moment by moment, "talking to unseen buddies, or react with various emotions to unseen events, without talking." In some cases consciousness was doubled: patients remained aware of the psychiatrist and spoke in the past tense. However the events were related, "the minuteness and wealth of detail which flood the memory, even of events which took place many months and even years before, is always impressive."[26] It was not just the detail but the "electrifying" intensity of the events that amazed

Grinker and Spiegel. They watched their patients' terror in "moments of supreme danger, such as during explosions within the plane, the falling of a plane, the mutilation or death of a friend before the flier's eyes."[27]

Whereas Sargant saw such productions as a mixture of fantasy and memory, Grinker and Spiegel baldly called them "true counterparts of what actually took place, rather than fantasies such as are produced in dreams or hypnotic states."[28] They even claimed that these acts of remembering could be *more* faithful to the historical past than the original experience, because soldiers had repressed their emotional responses initially, so that they experienced these emotions *for the first time* during therapy. "The emotional reactions (of the remembering patient) do not necessarily represent the actual behavior of the flier during the original episode, but rather what he repressed and controlled in order to carry on his job." So what the psychiatrist saw was, if anything, *more* true than both the soldier's conscious memory and his initial experience. This was because the sodium amytal trance brought back both the events (which the conscious memory had repressed) and the emotions (which had been repressed during the *original* experience). Treatment made it possible for the patient to bring them both "into a complete whole."[29]

Eventually the drug's effect would begin to wear off. At this point the patient entered a "twilight" in which he was "strongly in contact with painful combat situations and the feelings aroused by them" and also "in contact with his immediate environment, the therapist and the present reality."[30] This was the moment when it was possible to help the memory to become "integrated" into the patient's consciousness. The patient was kept awake, to ensure a smooth transition from reliving to ordinary recall.[31]

Why did Pentothal work? Grinker and Spiegel thought that the drugs broke a psychological "vicious circle" in which the ego was kept too weak by intense stress to develop appropriate perspective on the original stressful events. Under Pentothal, "the ego can afford to look squarely at the situation, in part at least, and to find a new strength." The beginning of this process was the "pharmacological effect of Pentothal on the diencephalons," the part of the brain that Grinker thought managed "emergency expressions." Barbiturates had a greater effect on the diencephalons than on the cerebral cortex, and this gave the cortex the freedom (while the diencephalons were suppressed) to "re-establish its inhibitory function." When the ego was no longer "harassed" by an overpowering anxiety, it could reestablish its "appraising

Patient falling under the influence of Sodium Pentothal. *Combat Exhaustion* (1945), U.S. Army Signal Corps.

Flashback representing the sudden return of the soldier's memory under Sodium Pentothal. *Combat Exhaustion* (1945), U.S. Army Signal Corps.

functions," and the individual became able to face the traumatic situation. This allowed the ego to "dominate the emotion" and "initiate new steps to control the situation." This, they concluded, "constitutes the benign circle."

On this account, intravenous barbiturates were indicated in all cases of severe anxiety—"mutism, stupor, regressive somatic manifestations, regressive . . . [and] psychological manifestations, and amnesia." Narcosynthesis topped the list of treatments. Only after considering it were psychiatrists to consider or add other protocols like brief psychotherapy, convulsive shock, continuous sleep, occupational therapy, or group therapy.[32] Grinker and Spiegel wrote up their techniques and findings in a series of medical articles and books. An initially classified document called *Men Under Stress* gave their most general account of trauma. At the end of the war, it was widely read and reprinted over several editions.

Memory on Film

Grinker and Spiegel's approach spread rapidly. It was widely used in the European theater, particularly by doctors with little psychiatric training. Its appeal lay partly in the fact that it was easy to administer by anyone with a basic medical background, because Pentothal was a staple intravenous anesthetic: government records confirm that Pentothal was seen as an expedient psychiatric resource because nonpsychiatrists already knew how to use it.[33] But its effects were straightforward and often dramatic. It produced an almost mechanical or surgical manifestation of memory, with no subtleties requiring expertise to interpret. By late 1943 each military unit carried drugs that included sodium amytal, Nembutal, and Pentothal ampoules, tablets, or capsules.[34]

Among the most influential means of establishing this practice were a small number of widely distributed motion picture films made by the U.S. Army Signal Corps to train field medics in how to diagnose and treat psychiatric disorders. *Combat Exhaustion* (1945) was one of these. It began by rejecting the idea that battle trauma stemmed from personal weakness or cowardice, or that sufferers were malingering. It then outlined a diagnostic and therapeutic framework for understanding trauma, dramatizing a visit by medical trainees to a psychiatric hospital. A guide explained that 20 percent of military casualties were entirely psychiatric and remarked, in a phrase common to the psychoanalytic understanding of battle stress, "All men have their breaking point." When soldiers were pushed past that point, they experienced symptoms of emotional exhaustion. Ideas that could not be admitted into consciousness were then expressed symbolically through psychosomatic problems, including amnesia or aphasia.

The trainees toured the wards, examining patients. The key ailment was a form of hysteria characterized by somatic complaints caused by a repressed memory. This was demonstrated in the film by a soldier who suffered from back pain and could not walk. The guide explained that doctors could find no somatic cause for his difficulties. Pentothal would help to bring back his memory—to integrate it into his conscious mind and sense of self. The lecturer showed how to measure the right dose, then how to deliver it.

Then the patient slipped into the trance, whereupon his memory returned, suddenly and violently. The film cut to an intense battle scene in which the panicked soldier was alone and expected to die. At the end of this several-minute scene he cried out that his back hurt, marking the onset of his hysterical problems. With the help of the doctor, he emerged from the Pentothal trance with his memory restored and soon discovered that his pain was gone. This was the climax of the film. With the patient's recovery, the presentation ended.

Films like this were widely circulated in each army. They played a prominent role in what little officers learned about psychiatric problems: frontline officers received only a few hours or days of training in how to handle traumatic neurosis, and in some cases such films supplied most of the "training" a medical officer might receive.[35] It was significant that *Combat Exhaustion* did not just film the patient describing his memory but used a flashback, thus allowing the audience the illusion of sharing in the experience. The technique partook of a representation of returning memories as being "filmlike."

During the war, the sudden return of memory came increasingly to be identified with the film technique of a "flashback"—a scene that rehearsed an act of recollection by a character. Flashbacks had been used in motion pictures for a generation, but they had a new significance in the psychiatric films of the 1940s, and in feature films with plots involving amnesia or battle exhaustion. Motion picture flashbacks and returning memory phenomena both involved a sudden shift from one narrative frame, one state of mind, to another. The flashback involved a "cut" from one scene to another; the abreaction involved a sudden change of mental state from the present to the past relived. Audiences would have understood that memories do not (ordinarily) unfold moment by moment in real time. But film flashbacks literally represented the patient's experiences. Flashback scenes delivered to viewers the real-time experiences that psychiatrists claimed were played out in a patient's consciousness.

According to military records, the skills and perceptions such films conveyed took hold across the military domain. Psychotropic practices were commonly used on a wide range of patients and applied repeatedly until the individual realized the experiences held no power to hurt him.[36] Repeated drug-facilitated remembering could transform toxic experiences into healthy memories, according to reports from various military hospitals. For instance, at the 130th General Hospital, Major Howard Fabing worked with patients suffering from "blast syndrome." Fabing found that through "Pentothal hypnosis" a patient who claimed to have been rendered unconscious for some time by a nearby explosion could regain his memory for the entire period. While the patient remained under the influence, Fabing would "suddenly and dramatically reproduce the situation with battle noises," to give the impression that he was still at the scene. The drugs did not necessarily produce an immediate cure; if the patient fell asleep at the end of the session and then woke naturally from the drugged state, "the cure would not be complete."[37] In such cases, some doctors tried repeated drug sessions, others used psychotherapy. At one hospital, patients were kept awake as the effects of the drug wore off, in order to "tie" the newly restored information into the patient's consciousness.[38] But Pentothal was nevertheless a central tool of the doctors.

This massive and urgent memory program was not without its critics. In Britain, M. B. Wright argued that the treatment was deceptive—and dangerous—because patients often relapsed, and their conditions were worse the second time around.[39] The influential American psychiatrist M. Ralph Kaufman thought Pentothal interfered with the development of a therapeutic relationship between psychiatrist and patient, and he worried too that drugs could "'fix' in place" amnesiac symptoms. Kaufman described the procedure in detail, in a way that made it clear that he was very experienced with it. Like Grinker and Spiegel, he explained it as a realistic reliving of an original event, though he was far less sanguine about its therapeutic power.[40] Yet the wartime consensus lay with Grinker and Spiegel, and with the rapid rehabilitation of the traumatized so that they could be returned to action.

Harvard anesthesiologist Henry Beecher later recalled the implications of this consensus. Looking back on a period of amazing revelation about what low doses of drugs could do to the brain, he remembered in particular the case of a soldier named Wyatt, who had been caught in an open field by an enemy tank. Through some "perverse whim" he was not shot but was toyed with by the tank crew. He threw himself into a slit trench, and the tank rolled over it without harming him. But then it "reversed and passed again over the trench," and he realized that "he was going to be ground under the tank treads." With each pass, the tank's "whirling cleats came nearer." Wyatt lost consciousness—or, Beecher speculated, "became amnesic." Eventually the tank gave up and Wyatt survived. Beecher met him some time later in a psychiatric hospital. "He remembered nothing of it and remained a shattered man, driven by nightmares, unfit for combat or for life outside a hospital." His case seemed hopeless. But then he was "put half under anesthesia and this blank period probed by a psychiatrist. Finally, he remembered and could talk about it." Beecher acknowledged the controversy about the longevity of the effects but found the doubts to be of secondary importance. "The immediate insight and information obtained and the immediate therapeutic effects are spectacular."[41]

Let There Be Light

At the end of the war, filmmaker John Huston received a commission to document the experience and the recovery of psychiatric casualties. He was to show that traumatized soldiers were ordinary people subjected to extraordinary stress and capable of a recovery that would allow them to rejoin society. The idea was that the power of film would reassure the public—and especially potential employers—that veterans were not "dangerous" or "permanently damaged" but were selves capable of full social existence.[42] Ironically, the technology of representation Huston adopted proved *too* effective. The film, *Let There Be Light*, was so powerful that the army decided to suppress it.

When Huston read about Pentothal and hypnosis, the information struck him with the power of "a religious experience," and he made hypnotic states central in his film.[43] He immersed himself in life at Mason General Hospital, even learning hypnotic techniques himself and standing in for the staff hypnotist when he was unavailable.[44] Huston shot the film entirely within the hospital, tracking patients over two months. He filmed each scene simply, using long takes. The result, in marked contrast to *Combat Exhaustion*, implied that patients themselves were telling their stories without interference, and that their experiences were being conveyed with direct, unedited immediacy and veracity.

The film gave several examples of therapy with Pentothal and hypnosis, each more dramatic than the last. Huston's scenes were more subtle than the contrived Pentothal interview in *Combat Exhaustion*, yet more detailed than those in Sargant's film about war neurosis. In basic terms, though, it resembled the two earlier films in narrative structure: the patient moved from one state to another—not just from a waking state to a trance, but from a dislocated state to an anchored one, a false to a true. The treatment was shown restoring something that was visibly lost or out of kilter to its normal functioning.

The first case shown was a man suffering from psychosomatic partial paralysis. Under Pentothal treatment he suddenly found that he could walk, though no new memories surfaced to explain his problems. Hypnosis helped a second man suffering complete amnesia: he was placed in the hypnotic state and then responded to questions about his battle experience. When asked his name, he supplied it, along with the names of his family. He was then roused from the trance. The scene ended with the camera pausing for a moment on his expression of relief and happiness.

The third case involved symptoms less disabling than the first two, but the results were dramatic. The patient suffered from hysterical stuttering. Almost immediately after receiving Pentothal, he cried out, "My God, I can talk! I can talk! I can talk!" After a few moments, he became calm enough to specify when his stuttering began. It started with a difficulty pronouncing the *s* sound—which he then associated with German artillery fire. A voice-over helpfully supplied a rapid, rhythmic sybilation, *ss-ss-ss-ss*). This last scene was one long, unbroken take, enhanced by music to accentuate climactic moments and a short explanatory narrative toward the end. The patient resembled the amnesiac in the Sargant film, but Huston eschewed Sargant's didactic before-and-after shots. The soldier's manner lacked the overt drama of the actor in *Combat Exhaustion*, but his shaking body and voice had an intensity that was perhaps more affecting.

Let There Be Light was immensely important to Huston, who described it as his most moving and joy-filled experience as a filmmaker.[45] He was confident that it delivered the humanizing account of war trauma that the army desired. But on the day of its release, the film was withdrawn by military officials, and it was suppressed for decades. It was released only in 1980, by an

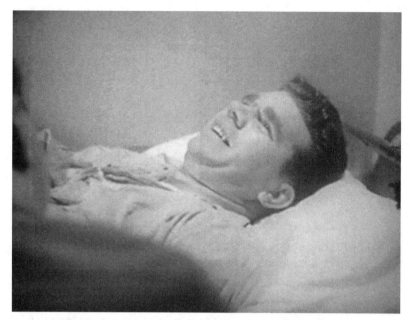

Stuttering patient recovers suddenly under the influence of Sodium Pentothal. *Let There Be Light* (1946), U.S. Army Signal Corps.

order from then vice president Walter Mondale. Huston himself thought it had been censored to keep it from undermining the army's "'warrior' myth, which said that our American soldiers went to war and came back all the stronger for the experience."[46] That comment itself was apparently made about 1980, when the military was being sharply criticized for taking this view of soldiers, and when the newly codified definition of posttraumatic stress disorder (PTSD) was much in the news. Yet a piece of evidence from the 1940s could support Huston's later interpretation: the army at that time displayed a similar though, as it proved, shorter-term reluctance to allow the publication of medical texts with messages similar to that of Huston's film. The unpublished text of Grinker and Spiegel's canonical work *Men Under Stress*, for instance, circulated among military psychiatrists and policy-makers for years before it was released as a widely published trade book, but typescripts in army archives are stamped "Secret"—a decision that seems bizarre given the widespread published literature on the subject. Huston's powerfully moving film, with its strong personal emphasis on the experience of individual soldiers and its intended civilian audience, would have seemed exponentially more risky than Grinker and Spiegel's accessible but rather dry tome.[47]

Adjudications

Before World War II, amytal and Pentothal had been associated with the controversial notion of a truth serum. As we have seen, the wartime memory therapies were thought to restore a subject's "true" self by bringing back to consciousness knowledge of some terrible experience. But another kind of truth was sought using these drugs, particularly in the second half of the war—one that resonated more with the earlier forensic associations. They could also be used to expose a "false" illness in a malingerer.

As barbiturate treatment became routine, doctors thought they knew what to expect when a truly afflicted person received the drugs. That made them increasingly likely to suggest other causes when the expected did *not* happen. For instance, London psychiatrists Carl Lambert and W. Linford Rees used barbiturates to treat amnesia and aphasia in Guy's Hospital and felt confident that they could identify malingerers: "Most patients become communicative under barbiturate narcosis," they remarked, but "the malingerer usually becomes more quiet and guarded."[48] When malingerers were asked questions during narcosis, they fell silent. Lambert and Rees's experience with the technique encouraged them to conclude that it was the patient, not the drug, that must be questioned. A true amnesiac would pour forth information in welcome relief.

Malingering was a serious management problem in the second half of the war, as the military tightened its conditions for psychiatric discharge. By late 1945 almost as many articles were being published on the use of barbiturates to detect malingering as on their use in neurosis. One reason was, perhaps, that the barbiturates helped to stabilize the diagnosis of neurosis itself. Neurosis had long been hard to pin down. Now, through training films like *Combat Exhaustion*, any doctor could identify and treat it. The training was supposed to make doctors be sensitive to neurotic symptoms rather than approaching all such cases with a suspicion that patients were merely cowards or frauds. But these simple training techniques taught doctors to expect a very specific range of reactions from traumatized soldiers who took Amytal or Pentothal. Neurosis came to be defined as the condition that was temporarily ameliorated by the application of sodium amytal. An inappropriate or adverse reaction to treatment suggested that no neurosis existed and that the patient was therefore "malingering."[49] Individuals suspected of feigning amnesia were described by some doctors who judged their cases as "refusing" to speak while under the influence of the drug.

The American military doctor Alfred O. Ludwig took this to its logical conclusion by publishing a formal technique for exposing malingerers. Anyone familiar with war neuroses would be "struck by the atypical and forced nature of the simulator's behavior," he wrote. "Soldiers with anxiety states show monotonously similar symptoms." They were restless, tense, oversensitive to light and noise, and reclusive. Malingerers, on the other hand, were exuberant, argumentative, and insubordinate. All of this was the stuff of impressions, but a clear distinction came with the drugs: "Under narcosis, [patients suffering from hysteria or anxiety] talk freely. . . . The amnesic material is usually recovered and the patient relives the traumatic episode with convincing realism. The malingerer resists narcosis, fearing that it will make him tell the truth. Narcotized, he fails to show any of the productivity of a neurotic patient and combats any effort to recover his lost memory with . . . negativism."[50]

One soldier claiming amnesia, Ludwig recounted, received Amytal twice, and both times "refused to answer questions at all or he answered with 'No more questions; don't ask any more questions.'" Ludwig decided he was a fraud. This soldier did eventually confess to malingering, though not until after he had been moved to two successive hospitals and had not been under Ludwig's care for some time.

Ludwig was one of many military doctors to use Amytal as a truth teller—not as a technique for making a suspect speak the truth, that is, but as a way to make his *behavior* betray the truth.[51] Indeed, Amytal was soon seen as better suited to the diagnosis of malingering than to the extraction of accurate narrative confessions. This is ironic, because what was being produced was no longer narrative truth, or indeed statements of any kind, but a form of withdrawal that was taken as a sign of concealment. After the war, Amytal was rarely used to produce confessions. Far more often it was employed as part of the evaluation of someone's behavior. The lack of a confessional state was confession in itself.

Full Circle: The Return of the Truth Drugs

Toward the end of the war, the idea that drugs could reveal the deceits of suspected malingerers gave way in turn to other forensic ambitions—ambitions that had much in common with the drugs' early history as supposed agents of "truth." The idea of a truth drug had been mooted again—and pursued in a speculative way—in military intelligence research carried out by the American Office of Strategic Services (OSS) in 1941 and then in 1943–44. The OSS's purpose was to interrogate prisoners of war who might "possess knowledge of particularly important matters" and to evaluate Americans "for suitability for access to the highest level secrets," by making the individual "uninhibitedly and irresponsibly loquacious in a way that was harmless, undetectable, and left no memory of the event."[52] The old truth drugs were found ineffective, as were many other psychotropic drugs. But one new compound, tetrahydrocannabinol acetate, an extract of marijuana, was found to work. It was successfully used on a New York gangster, but its broader use was ruled out for political reasons.

The ambition for a truth drug had clearly not gone away, and barbiturates soon found a career for this purpose outside military intelligence. Their potential for exposing malingers revived their associations to "truth." There was a revival of similar enterprises, this time involving a blend of psychiatric evaluation (for the insanity plea and for competence to stand trial) with the kind of forensic questioning that had been associated with truth serum before the war. Several conflicting notions of memory and of "truth" became entangled.

A good example of how this happened is the case of William Heirens. In 1946, Heirens was a seventeen-year-old first-year student at the University of Chicago who moonlighted at housebreaking. He was arrested in the course of one of several burglaries. But he soon came under suspicion not just of burglary, but of serial murder. Police claimed to have found his fingerprint on a document associated with one of a set of killings they had been investigating. Under interrogation, Heirens claimed to have lost consciousness when he entered the apartments he had robbed. He also claimed to have gaps in his memories, and he even said he had lost consciousness during the interrogation itself. Roy Grinker and another psychiatrist, William H. Haines, were asked by state's attorney William Tuohy to do a Pentothal interview to determine whether Heirens was faking his claims of altered consciousness and amnesia.

Haines, a government employee whose job was to make competency evaluations, had never once made a diagnosis of insanity.[53] He concluded that Heirens was malingering to avoid taking responsibility for the crime. His interest in using Pentothal was never clearly articulated, but it is likely that it followed from tests like those described by Ludwig just two or three years earlier, which were premised on the idea that Pentothal could help one tell the difference between a real memory problem and a fraud. Grinker, on the other hand, saw Pentothal as a true memory refresher and found the claim of dissociation more plausible than Haines did.[54] For him the Pentothal interview would tap Heirens's memory and could provide evidence of incompetency, because Heirens would not have been able to judge or to control his actions. But Heirens did not acknowledge any part in the murder during the Pentothal interview.[55] Grinker concluded that he was a "dissociated, psychotic schizophrenic"—a characterization that one psychiatrist who knew Grinker thought was designed to help him make an affirmative defense of insanity. But he drew no conclusions from the interview about Heirens's actions.[56] Authorities later claimed that Heirens accused an alter ego, George Murman, of the crime, but what he actually said is in dispute, and records of the Pentothal interview have disappeared.

But if Heirens made no confession, his supporters have claimed over the years that the session did edge him toward a guilty plea. It was part of a string of aggressive tests, carried out over several days without legal representation or the support of his parents. Meanwhile, word in the press was that he had indeed confessed, and Heirens's lawyers persuaded him to enter a guilty plea in order to avoid the death penalty. He received three consecutive life sentences. Heirens retracted his confession shortly after sentencing, claiming he had made his statements under duress, both from the police and in the form of aggressive advice by his attorney, in an attempt to avoid the electric chair.[57] As of 2011 he is still in jail, as Illinois's longest-serving offender.

Grinker's and Haines's respective motivations for trying Pentothal—in Grinker's case to bring back a blocked memory, in Haines's to expose "malingering"—exemplified the two conventions that came out of the war. The extension of Pentothal treatments into civilian contexts to evaluate what the military called "malingering" was part of a kind of diagnostic "creep" of its perceived utility and legitimacy. That is, in Heirens's case, some of the parties hoped to achieve more than just knowledge of whether Heirens's loss of consciousness on arrest had been feigned: they wanted, as prosecuting attorney Tuohy said, to "get the truth out of him."

When Grinker and Spiegel inserted an IV into a patient for the first time to revive blocked memories, they were creating a variant on an old psychoanalytic practice: the review and discussion of patients' memories. Classical psychoanalytic theory stipulated that it did not matter whether the memories were actually "true"; analysts expected that what patients recounted would be affected by their life in the present. But this did not matter as long as a patient believed they were true, and as long as the event recounted was emotionally significant. Grinker and Spiegel would have been trained to take this approach to their patients' acts of remembering. But the exigencies of the war encouraged a more mechanical understanding. Missing memories were now like missing puzzle pieces, which could be snapped back into place by pharmaceutical levers. This in turn led to the idea that simple historical "truth" itself was a characteristic of some kind of raw memory record that could sometimes be retrieved.

The most general result of these wartime therapies was a broad dissemination of this mechanical idea of veridical memory. It came to pervade psychiatry and public discussion alike. During the war, thousands of doctors and tens of thousands of patients were introduced to psychiatry through these memory techniques. In the first years after the conflict, this understanding of memory moved into civilian culture. Army doctors took up civilian practices and, in record numbers, got formal qualifications in psychiatry and psychoanalysis. Feature films made flashbacks and abreaction central to their plots. And new medical, surgical, and psychological discoveries seemed to confirm and extend the idea of movielike memories secreted in the subconscious.

attendant emotion."[3] These records were not ordinary memories, however; their purpose was to provide information for other functions, like recognition. Only in unusual circumstances, like his operations, were they actually revived and displayed to the self. Even if these were not memories per se, Penfield therefore conceived that studying them could unlock "the nature and the point of impact of the nerve impulses" that were involved in summoning an ordinary memory in the normal act of remembering.[4]

The Stream of Consciousness and the Physiology of Time

Penfield's findings formed a physiology of time itself. The past became part of the body, where it lay preserved for future reference. He wrote that the "thread of continuity" in these memories seemed to be time: "The original pattern was laid down in temporal succession. It is the thread of temporal succession that later seems to hold the elements of evoked recollections together."[5]

Penfield's stimulus produced a "picture," but not a static one. The recalled images and sounds were as dynamic as they had been as original experiences. Penfield concluded that "the memory record continues intact even after the subject's ability to recall it disappears," and the extraordinary situation of his surgeries produced recollections that retained "the detailed character of the original experience." They could be so lifelike that they appeared "to be a present experience."

Patients' own narratives, recorded during the operations, substantiated this claim. In the case of D.F., "A point on the superior surface of the right temporal lobe was stimulated within the fissure of Sylvius, [and] the patient heard a specific popular song being played as though by an orchestra. Repeated stimulations reproduced the same music. While the electrode was kept in place, she hummed the tune, chorus, and verse, thus accompanying the music she heard." Another patient reported,

> "My mother is telling my brother he has got his coat on backwards, I can just hear them." He was asked whether he remembered this happening. "Oh, yes," he replied, "just before I came here." He was then asked whether these things were like dreams, and he replied, "No . . . It is just like I go into a daze." . . . This boy of 12 was an accurate witness. Every effort was made to mislead him by stimulations without warnings and warnings without stimulations, but at no time could he be deceived. When in doubt, he asked thoughtfully to have the stimulation repeated before committing himself to a reply![6]

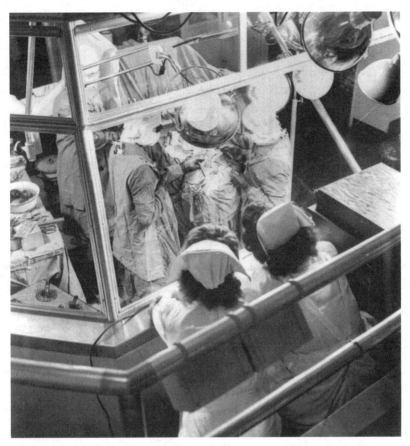

Operating theater and spectator gallery designed by Wilder Penfield, Montreal Neurological Institute. With permission of the Wilder Penfield Archive.

and their own ways of understanding psychic phenomena consequently had a far greater effect on him than one might have expected from any mid-twentieth-century doctor, much less a surgeon. For most surgeons, whose ideal subjects were pliant bodies passively awaiting the knife, it was disturbing to encounter a technique that depended on a patient's contributions. One of Penfield's colleagues has told me that many other surgeons were daunted by the thought of such a relationship.[9]

These collaborations were central to Penfield's work on functional localization. He often stressed the patient's ability to make reliable distinctions between occurrences of different kinds—for instance, between voluntary movements and those produced surgically by electrical stimulation of the motor cortex. The confidence he had not only in his own analytical powers but also in those of his patients gave credit to their extraordinary claims that these were not hallucinations that merely *appeared* to be memories but were instead authentic events from their pasts. His own view of their experiences evolved from hallucinations that were *like* memories to actual experiences stored perfectly in the brain, which were being revived in the operating theater. In some patients, he announced, stimulation of the temporal cortex could "awaken a memory or cause the patient to experience a dream that is made up of materials from the storehouse of his memory." For instance, stimulation might cause a patient to hear a familiar song. "This is not a static impression like a picture," Penfield maintained, but "progressive." As long as the electrode remained in place, "the tune may pass from verse to chorus."

Penfield considered his patients to be fellow explorers of the great undiscovered country of the brain, and together they built the maps for which he has become famous. He described things in language that echoed his patients' categories, even though his own descriptions used neurological knowledge they could not have shared. For instance, his patients likened their experiences to motion pictures before he did. Penfield's approach therefore combined the popular with the elite or scientific, the lay with the specialist. It was this unusual characteristic—the authority of the specialist neurosurgeon combined with the accessibility of lay categories of explanation—that made his claims so powerful in midcentury North America.

Neurosurgery as Psychological Experiment

As his interest in memory grew, Penfield turned to a community that had never interested him before: psychologists. His reflections, leading as they did toward a claim that memory traces were permanent and lodged locally in neuron trails, could not have conflicted more with the new claims coming out of experimental psychology. The experimental psychologists tended to favor a view that memories were in constant flux, and that they consisted not in highly localized neuronal changes but in patterns, or networks, of neuronal facilitation. Frederic Bartlett's constructive account of memory, discussed in chapter 9, of course spoke against the principle of permanence. Bartlett explicitly rejected the idea of stable memory units stored in the brain, and his experiments portrayed memories in a dynamic relationship with future experiences. His memory work of the 1910s and 1920s had laid a foundation for researchers across many fields of the social sciences to develop a "constructivist" account of memory, meaning, and knowledge.

One salient line of research was based not on human behavior but on animal behavior. Karl Lashley had spent the 1930s and 1940s first at the University of Chicago and then at Harvard in pursuit of what he called the "engram"—a physical location and structure embodying discrete memories. Lashley trained rats in mazes, then tried to destroy their memories surgically. It was a total failure. He could not identify a specific area of brain tissue linked to rats' ability to run their maze. The more brain tissue he removed, the less functioning remained, but only the quantity mattered, not its location. Lashley concluded that memories were not localized. It might be appealing to think of memory "units" stored in specific locations in the brain, he decided, but in fact there were no such things.[10]

Penfield would have known of Lashley's work from Donald Hebb, who held a postdoctoral position at the MNI in the late 1930s and remained in Montreal after that. Hebb found evidence to support Lashley's distributed model of mind. The abandonment of the idea of a straightforward location for memory traces would later be seen as a significant step in the development of neural-net theory, which Hebb pioneered in the 1950s.[11] Hebb developed a synaptic account of learning, based on the relation between the frequency and quantity of neuron firings and the growth and efficiency of the synapse. He also developed the notion of "cell assembly," a progenitor of modern neural net conceptions. In 1949 he published his most influential book, *The Organization of Behavior: A Neuropsychological Theory*, a work that placed McGill on the map of the neurosciences. Hebb's work soon made McGill a major center for neuropsychology.

The idea that a surgeon could stimulate brain cells where certain memories were stored and call them back to life was therefore a particularly controversial proposition. Penfield began to make it just as Lashley was moving in the opposite direction by publicly abandoning his "search for the engram."[12] Yet in the same few years, Penfield not only began to air his claims about memory but even started publishing rough maps of cerebral locations. Stimulation at a single point in the brain could, he argued, activate a "progressive state which is in a series of neurones." Expressed in this way, his work represented a fundamental challenge to several research traditions at once.

Hebb's work had meanwhile attracted the attention of a woman who would become one of his most famous students, and the most important of

Penfield's collaborators. Brenda Milner had come to Montreal only a few years earlier from Cambridge, England, where she had worked in Frederic Bartlett's laboratory. She moved to Montreal with her husband in the late 1940s. There she found a job in the psychology department at the University of Montreal. But the environment was not congenial to this experimentally oriented Anglican psychologist steeped in the British empiricist tradition. As she later put it, the University of Montreal was "very psychoanalytic and very Catholic."[13] She was the only faculty member teaching experimental and comparative psychology. Meanwhile, McGill was trying to build up its psychology department, which was in the "doldrums" after the war. Hebb had been hired in an effort to jump-start it. Milner, who had not finished a PhD during her time at Cambridge, now decided to undertake one with Hebb as her adviser. He suggested she do some work at MNI, observing Penfield's patients.

Milner was not an obvious choice in some respects. She had no medical background, and no experience in physiological or neurological laboratory work. Even more problematic was the stamp of her previous teacher, Bartlett, from whom she took an interest in how the mind actively constructs "meaning" and an assumption that memories change over time. She had not studied memory during her time in Cambridge, but her first work in Canada centrally involved teaching Bartlett's theory of memory. Bartlett's goals and methods were worlds away from the stimulus-response experiments in vogue in North America at this time, which were more resonant with Penfield's brain-probing electrical techniques. But Milner found herself fascinated by Penfield's patients and decided to concentrate on the psychological issues they raised. Hebb discouraged the plan, worried that Penfield would never accept Milner's psychological approach. Nevertheless, Milner was determined to persevere, and she settled into work at MNI in 1952–53.

That very year, a series of disastrous cases placed the field of psychology front and center of Penfield's concerns. Penfield had thought that psychology was common sense and that he did not need expert help in the area. All that changed in 1952, when Milner was making intensive psychological studies of patients before and after surgical excisions had been made to the temporal cortex. By this time Penfield's operations had become highly invasive. In his initial work, he had hoped that small excisions on or near the surface of the brain would be sufficient to ameliorate a patient's problems. He would remove a superficial area of the cortex, after using his "Montreal technique" to find the point from which the seizures were arising. But this approach did not work: the surrounding tissue later became very active, and patients continued to have seizures. By 1950 he was going deeper—delving beyond the cortex to excise parts of the amygdala and hippocampus. This more invasive approach supplied the conditions for a series of disastrous operations that dramatically changed the way Penfield thought about memory.

In 1952–53 Penfield and Milner encountered what Milner later called "two horrible cases" that threw their enterprise into crisis. The first case, a patient known as P.B., was an engineer from the New Jersey. He had had a previous brain operation at the cortex, but then he suffered recurrent seizures. He later returned to MNI to have what had become a standard procedure in such cases, the removal of the amygdala and part of the hippocampus. Before the operation, his memory was fine; after it, he suffered severe memory loss. Weeks later, Penfield had another case of temporal lobectomy, which left the patient with truly devastating memory problems. What was particularly terrible about this second case was that it the procedure was not a matter of life or death but an elective surgery undertaken in a bid to reduce the patient's seizures. Milner and Penfield later concluded that there must have been something wrong on the other side of the brain; when Penfield operated, the patient was left with no ability to store new memories on either side. This speculation was confirmed, years later, by autopsy. But this scarcely reduced the disquiet the event provoked at the time.

Milner and Penfield noticed that removing certain parts of one hippocampus resulted in the loss of short-term memory in patients who had had preexisting damage to the hippocampus on the other side. Such patients became unable to add new memories to their long-term store of information, though they did retain their existing long-term memories. This meant they could not remember what had happened just a few moments before the present. This was devastating for them and their families. It was also distressing to Penfield and his collaborators, whose work relied on the symmetrical structure of the brain to ensure against such profound losses. Their operations for temporal lobe epilepsy were always done on one lobe of the brain, on the assumption that the other side was functional. Before the age of functional imaging, however, there was no way to confirm this clinically. Even the introduction of the EEG did not provide definitive reassurance on this point, because such electrical recordings were thought to be highly ambiguous.

When Milner spoke about the two cases at a conference in 1954, she met American neurosurgeon William Beecher Scoville. Scoville was struck by similarities between Milner's cases and one of his own, "H.M."[14] Henry M.

had suffered from an extreme form of epilepsy that grew progressively worse until he was no longer able to work. Scoville was experimenting with surgery as a treatment for psychosis, and H.M. was referred to him in the hope that neurosurgery could cure the seizures. Scoville determined that his seizures were local to the temporal lobe, and he decided to remove large portions of both temporal lobes. The operation excised all of the amygdala, the perirhinal cortices, and a large part of the hippocampus—a very dramatic intervention indeed. It did succeed in reducing H.M.'s seizures. But it also destroyed his existing memories and his ability to create new ones. For the rest of his life, he would be unable to retain any new knowledge for more than a few minutes, to identify individuals he met, or even to retain consciousness of skills he had attained.

Milner traveled to Connecticut to study H.M., who soon became one of the most famous cases in the history of neurosurgery—the focus of hundreds of research papers and projects and the engrossing subject of Milner's own research for years to come.[15] Milner's studies showed that, memory aside, in many other respects he remained neurologically and cognitively competent. He performed above average on IQ tests and was capable of learning several skills (though not of remembering that he had learned them). Before his death in 2008 he lived in eastern Connecticut, the subject of continuous psychological and neurological study. Studies of H.M. have continued even after his death in December 2008. His brain was taken to the University of California at San Diego, where it was sliced into large histological sections that are now the subject of further research and are available online.[16]

Penfield and Milner's two "horrible" cases of 1952–53, along with the case of H.M., launched decades of research into the neuropsychology, physiology, and anatomy of higher brain function. This work suggested that the hippocampus was important to storage and recall. But the cases also indicated that some other part of the brain must be the repository of long-term memories, since these were unaffected by the loss of tissue in the hippocampus. Milner's work transformed the status of memory in the neurosciences by showing that both the creation of memory and stored information were localized in such specific parts of the brain.

Penfield and another colleague, William Feindel, made a second important observation during the early 1950s. This related to automatism, a detached behavior associated with seizures of the temporal lobe. They found that stimulating the amygdala produced artificial automatism, along with amnesia for the same period. Milner concluded that the brain had stopped recording experiences during this period and inferred, further, that this area of the brain was important in making a record of lived experience. During these events, they claimed, patients were no longer able to form memories from their experiences, and later they would have amnesia for the relevant period. As Feindel later put it, "In the tape recorder analogy, the recorder was not functioning during this time."[17]

As Penfield and his colleagues pursued these questions, Penfield's papers became increasingly confident in their claims about the "ganglionic record" of experience. In 1952 he named the "memory cortex" and presented it to colleagues as an unexplored territory, inviting researchers into a new "chamber of understanding."[18] The discovery that "cortical 'patterns' . . . preserve the detail of current experience, as though in a library of many volumes," would lead to a new "physiology of the mind." Penfield worked toward an account of a "neurone system centrally placed . . . [in the brain stem] and equally connected with the two hemispheres."[19] Now he had to study "the nature of the pattern, the mechanism of its formation, the mechanism of its subsequent utilization, and the integrative processes that form the substratum of consciousness."[20]

Yet Penfield's convictions retained an idiosyncratic character that set his claims apart from his psychological colleagues' ideas. Hebb's "cell assembly" was directly opposed to the idea that a memory could be located in a single neuron or even an orderly sequence of them. Even Milner flatly rejected the idea of memory as a stable archive. So although Penfield's work seemed to take him in a psychological direction, he insisted that this did not make him a fellow traveler.

Proust on the Operating Table

Penfield's work was exciting to another group of experts on the mind, one even more distant from the world of neurosurgery than the experimental psychologists: psychoanalysts. Surgeons and psychoanalysts were poles apart, the knife and the conversation being such different tools for excavating the mind. But the "memory therapies" developed by analysts in World War II had made them both interested in the idea of buried, intact records of past experience. Penfield's surgical results were tantalizingly evocative of the psychological mechanism of abreaction, in which past events seemed to be revived and re-experienced in real time.

Psychoanalysis was the dominant framework for mid-twentieth-century psychiatry. But it was not rooted in conventional medical research, its scientific authority was contested at best, and the causes it proposed for mental occurrences were mental themselves. Even its "data" were problematic from a neurologist's point of view, since they consisted mainly of conversations. Clinical psychologists were perhaps even worse, since they typically had no medical degree. Penfield himself had previously had no time for them. And psychoanalysts were not in the business of speculating about where memories were located in the brain. Their "maps" were metaphorical, representing dynamic relationships they perceived between different psychic processes. On the other hand, however, analysts had documented effects that could be seen as similar to Penfield's—the revival and re-experience of past events. Even if their explanations remained entirely alien to a neurosurgeon's approach and concerns, Penfield's memory patients had by now caused him to take notice of them. No wonder, then, that psychoanalysts were intrigued by Penfield's work. It had much to offer them, in terms of professional authority as well as psychiatric insight.[21]

Psychoanalysts' extensive experience with abreaction, particularly in the war years, primed them for the idea that raw experiences from the past could be reawakened under special conditions. But they had not attempted a physioanatomical account of *why* or *how* pent-up experiences could be replayed. Penfield's work offered one. As one excited acolyte proclaimed, it placed "Proust on the operating table." This exclamation came from Lawrence Kubie, an influential analyst at Yale. Kubie spent time in Penfield's operating theater, eventually staging free-association experiments on the operating table. At first he merely observed operations. But then he asked to participate. Scrubbed and masked, he would sit "under the drapes," the cloth barrier between the back and front of the patient's head. On one side of the drapes, the surgeon worked; on the other, the conscious patient could speak with an interlocutor who sat beside him. Kubie would place his Dictaphone under these drapes to record patients' statements during the electrical stimulation.[22] It was based on this experience that Kubie could announce that the memory effects Penfield produced established "that the remote past can be as vivid as the present." They demonstrated that "time and space offer no obstacles to unconscious processes" and showed "the precision with which the past is permanently stored as discrete units." The surgeon's knife was the new royal road to the unconscious.

Ever since its early days, proponents of psychoanalysis had been defensive about the biological underpinnings of their field's signature theoretical

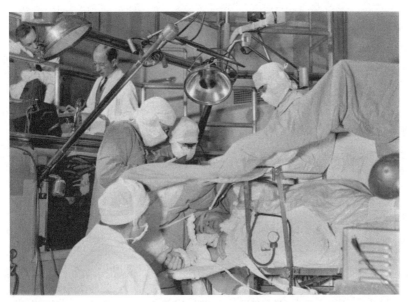

Operating room setup for the Montreal technique. With permission of the Wilder Penfield Archive.

entities. They had suffered the taunts of scientists who derided concepts like repression and the unconscious as vague and unproven. But now a brain surgeon with no psychiatric background had demonstrated the existence of "this 'library of many volumes,' unconscious yet dynamically charged with the lifelike vividness of sensation and affect." Kubie anticipated that as a direct result, "the fears of our more psychophobic colleagues should at last be set at rest." Penfield's work would reassure neurophysiologists who were still inclined to "boggle" over the term "unconscious" as applied to psychological processes. It confirmed the existence of a system that analysts had always had to resort to metaphors to describe and also offered a physiology of it.

The benefits of a collaboration between neurosurgery and psychoanalysis were not to run in just one direction. Kubie thought that psychoanalysis could offer something in return. When he heard of the terrible case of P.B., the patient who had lost all his long-term memories after removal of the hippocampus, Kubie refused to accept that the damage was permanent. After years of promoting barbiturates in cases of traumatic amnesia, he could not believe that the memories (and the capacity for creating memory) were permanently gone. He suggested an intravenous injection of Sodium Pentothal, which he expected to bring back the missing memories. Penfield and Milner thought the idea was ridiculous. They were astounded at the proposition that barbiturates could help a condition they thought was clearly the result of the removal of a major anatomical structure. Kubie's optimism was a sign of the extent of his investment not only in the use of Pentothal as a memory technique, but in the broader understanding of memory that was linked to it. He was permitted to try the Pentothal treatment. To his disappointment, it proved completely ineffective: P.B. simply became "rambling and foolish."[23] After this abject failure, Kubie proposed no further barbiturate treatments.

Yet Kubie's fascination with Penfield's approach persisted. He continued to attend operations, and in a series of papers and talks he elaborated on their broad implications for analysts. The cortical "patterns" that preserved records of experience became conscious after they became connected to "one or another of the indispensable substrata of consciousness, one of which may prove to be the centrencephalic system." This was to be thought of as the highest "level of integration" in the nervous system and was "comparable to the release or inhibition of a spinal reflex under the influence of upper motor neurons." There were, he thought, at least three levels of awareness, in which neurological structures could now be found corresponding to psychoanalytic categories: "three comparable relationships between the temporal cortex and the 'indispensable substrata' (centrencephalic system) are at least conceivable to correspond respectively to repressed (Ucs), released conscious (Cs), and augmented preconscious (Pcs) levels of function."[24]

It was in this sense that Penfield's effects reminded Kubie of "an electrically-stimulated *Recherche du temps perdu*." The difference, of course, was that the *temps* was not really *perdu* at all. If psychoanalysts and neuro-surgeons could collaborate, they might hope to learn "whether fully repressed experiences can be evoked electrically and relived with full somatic participa-tion, and then to discover whether electrically recovered memories are subse-quently repressed, if so what the ultimate effects of their electrical evocation may be, and whether such evocation is temporary or lasting."[25] Analysts and neurologists needed each other because these mechanisms were critical for understanding "both the psychodynamics of repression, and also its neuro-physiological substratum."

Analysts were indeed intrigued by these possibilities. The influential psy-chiatrist Sydney Margolin, for example, declared that his suggestions might lead to the discovery of a physiological substructure for psychoanalysis it-self.[26] And John Lilly, a biophysicist at University of Pennsylvania, expressed his own confidence that Penfield's patients were indeed "reliving and re-observing" past events—"directly, without the use of words." Lilly added that the work promised a "real fusion of psychoanalytic and neurophysiological research and theory." But Mortimer Ostow, another analyst, was more cautious, wanting to make sure Kubie was not claiming that Penfield could "evoke unconscious memories"—that "the unconscious memories which we labor so hard in analysis to evoke are laid down in the temporal lobe." Ostow's objection was based in part on the content of the memories that Penfield was able to secure. They did not include anything involving "a naked genital or pregenital fantasy," the meat and drink of repressed memories that psychoanalysts sought to explore. Neither were they emotionally significant events. The well-known analyst Ernst Kris was also unsure whether the memories involved repressed material or merely events that had not been reflected on for some time. But he too called for collaboration with neurosurgeons. "How do such imprints arise?" he asked; "How does the brain produce them in the mind?" Collaboration between analysts and neurosurgeons might one day make it possible to "express the psychic apparatus in concrete, neuropsychological terms." Several psychoanalytic papers did attempt to cash out the implications of Penfield's practices, but in practice such possibilities proved very hard to realize.

Memories on Record, Tape, and Film

For Penfield, memory was perfectly preserved in an orderly ganglionic record. It could be recalled from there and replayed under certain circumstances. Human memory recapitulated, in fact, the properties of the recording technologies—tapes, LPs, and film—that were set to characterize the cultural life of the postwar years.

Although he was surrounded by some of the brightest stars of experimental psychology and neurology, Penfield himself reached outside the academy for ways to think about memory records. He looked in particular to popular new media technologies to help him develop an account of "recording" that would capture his notions of preserved time and sequence. Initially, he reached for a familiar early twentieth-century technology of wire recording. But he quickly moved on to the more timely technologies of motion picture film and tape recording. He alluded to them hesitantly at first, but they took on an increasingly prominent role in his accounts during the next two decades.

In 1954, about to engage psychologists for the first time at an international psychology conference, Penfield prepared by dusting off his college copy of William James's *Principles of Psychology*. He found himself newly riveted by the chapter titled "The Stream of Thought." He felt that his memory effects were dramatic displays of the mechanism by which this "stream" was preserved.

> Among the millions and millions of nerve cells that clothe certain parts of the temporal lobe on each side, there runs a thread. It is the thread of time, the thread that has run through each succeeding wakeful hour of the individual's past life. Think of this thread, if you like, as a pathway through an unending sequence of nerve cells, nerve fibers and synapses. It is a pathway which can be followed again because of the continuing facilitation that has been created in the cell contacts.
>
> The stream of consciousness flows inexorably onward, as described in the words of William James. But, unlike a river, it leaves behind it a permanent record that runs, no doubt, like a thread along a pathway of ganglionic and synaptic facilitations in the brain. This pathway is located partly or wholly in the temporal lobes.[27]

Unlike James's "stream," the neurological "strip" of time could be replayed when "the neurosurgeon's electrode activates some portion of [the neuronal] thread." The brain responded "as though that thread were a wire recorder or a strip of cinematographic film." The difference between brain and motion pictures was that while each had a playback mechanism, the brain produced no special effects. Its film "runs forward, never backward, even when resurrected from the past," and it could neither be speeded up nor played in slow motion. It moved "at time's own unchanged pace." Indeed, once revived it had a power of its own: "There is no holding it still, no turning it back" as long as the electrode remained in place.

Penfield suspected that when these records were formed in the brain, they already contained "the elements of consciousness." He imagined that as each record took shape, "like a film, its contents are projected on the screen of man's awareness before it is replaced by subsequent experience." This record was thus more than just stored information: it was a final stage "in the neuronal integration which makes consciousness what it is."[28] He justified this centrality of consciousness to the memory record partly by a simple distinction: only "things of which the individual was once aware" were recorded. Absent from the record were "the sensory impulses he ignored, the talk he did not heed." He elaborated on this model over several decades, adding detail and insisting ever more emphatically on the permanence and authenticity of the neural memory record. In 1969, for example, he could still be heard affirming that "flashback strips of experience" of the kind his probes awakened included "all of the individual's awareness and nothing of what he ignored." The record was "a trail of facilitation of neuronal connections that can be followed again by an electric current many years later with no loss of detail, as though a tape recorder had been receiving it all."[29]

This distinction, so intuitive in Penfield's writing, would not seem nearly so simple to psychologists. Neither analysts nor experimenters made such hard and fast distinctions between attention and inattention. Analysts were more interested in mental and perceptual processes that operated below the level of full consciousness; and the beginnings of a consensus were already emerging from academic psychology laboratories on the existence of "subliminal" perceptions, which would grow into a fashionable research area in the late 1950s and 1960s. But for Penfield the distinction made eminent sense. One reason was that he conceived of the memory record in terms of cinematic film.

By the late 1950s, after likening memory variously to wire recordings and audiotape, Penfield was settling on motion picture film as his medium of choice. This analogy may sound more than a little naive, particularly for someone who filmed and photographed some of his own surgical operations. He would have known that any motion picture film is edited and that raw films do not display anything equivalent to the conscious experience felt by someone present during events recorded by the camera's "eye." And the technological differences between the recording techniques he invoked (wire recording, tape, motion pictures) similarly retreated into insignificance. Yet it seems that few if any commentators at the time criticized his accounts on such grounds. This makes sense if we consider the cultural status of motion pictures in the 1950s—particularly the films that were being made every day in homes across the United States.

Amateur film became a significant practice in the 1930s, when the first affordable movie cameras appeared on the market. It became more widespread in the 1950s. In that midcentury generation, cameras became cheaper still; photography journals began running articles and advertisements specifically aimed at the amateur movie maker; general interest magazines published articles on what would later be called home movies; and amateur film clubs proliferated. Amateur cameras were consistently marketed as an external

Vacation movies... so gloriously colorful

...so easy to make and afford

That wonderful trip can last for years—for today movies are so inexpensive they're "part of life" in a million-and-a-half American homes.

Faraway places . . . happy yesterdays . . . spring to joyous life whenever you wish, in movies you make yourself.

Advertisement for Kodak film, 1951.

form of family memory—and memory of a particularly lifelike kind. They were advertised as offering users the chance to relive special experiences over and over again. An article in the *American Photographic Annual* in 1936 stated bluntly that "films are memories in tangible form."

Domestic films preserved life intact, to be relived later. By the fifties they could even include a soundtrack. Home movie makers thus understood what they were doing differently from the approach of Hollywood directors and academic theorists. Münsterberg had once described film as an externalized form of consciousness that intervened in the viewer's experience. In contrast, amateurs saw their lightly edited (or unedited) film sequences as little slices of life, recorded without editorial intervention or hindsight.

A new "realism" was also a feature of other 1940s and 1950s film practices. Many projects tried to convey a sense of the "real," from gripping documentaries to the use of documentary footage or nonactors in fiction films. The innovation of Cinemascope, for example, was designed to fill the field of vision, so that one lost the sense of looking at something framed by a proscenium; the impression of being immersed in a visual-auditory experience offered a new convention for technological "high fidelity." Hollywood experimented with 3D in 1952. And the use of flashbacks as a film device peaked during these years, so that this kind of memory event became part of the distinctive character of film at this time. There were many ways a rising realist epistemology became attached to film.

Between home and Hollywood, film entered many different spaces in these years. After the war, the same cameras and assumptions traveled to new venues such as industrial or scientific sites, and cheap projectors that had been mass-produced during the conflict went out to public schools, where movies quickly became a familiar teaching tool. In all of these sites, film was regarded as a straightforward record of external objects or events, not a crafted means of intervening in a spectator's experience.[30]

The late 1940s and 1950s also saw an unprecedented influx of motion picture cameras in scientific and medical research. Indeed, surgeons were among the professionals that Kodak particularly hoped to reach. The company placed ads in medical and surgical journals, arguing that still and motion picture photography allowed a truthfulness in the documentation of an operation that was unmatched by other forms of record keeping. It is clear from the kinds of images they showed, however, that conveying subjective,

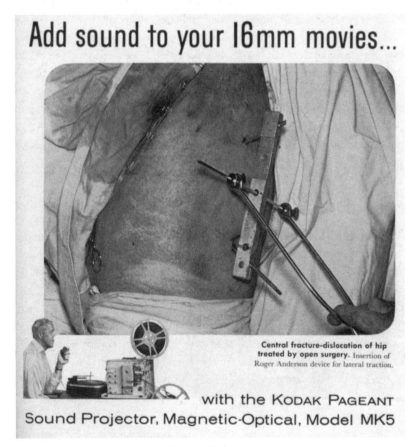

Add sound to your 16mm movies...

Central fracture-dislocation of hip treated by open surgery. Insertion of Roger Anderson device for lateral traction.

with the KODAK PAGEANT Sound Projector, Magnetic-Optical, Model MK5

Advertisement for a sound projector, Eastman Kodak, ca. 1950.

autobiographical narratives (whether of the surgeon or of the patient) was not the purpose. And although Penfield's own use of film was close in some respects to the representation of the surgical recording seen in such advertisements, this was not how he thought about film when he was listening to his memory patients. His understanding more closely resembled the domestic approach to movies. The idea of "living it over again" was central, and that idea chimed very well with the realistic immediacy patients themselves reported.

Journalists were one group who found the notion of a neural movie particularly appealing. They wrote in awe about the power of electrical stimulation to "recreate an entire segment of a person's past life, complete with sight, sound, light, color, dimension, smell and the appropriate emotional feeling."[31] *Time* described the brain as "an audio-visual tape recorder, preserving the details of everything a man sees and hears." Press accounts like these tended to evoke expensive, and indeed rather exclusive, pieces of technology—the high-fidelity stereo, the home cinema, the top-of-the-line home movie camera. But above all they implied a common ground between what one might call hardware and wetware, machines and brains, as memory devices. Moreover, they suggested that wetware was a kind of personal commodity. It was as if life itself became personal capital, to be stored up by neural recording devices and then drawn on and used at a later point. All of us, it seemed, not only carried a "built-in hi-fi," but acted as its pressing plants and customers too.[32]

The Art of Memories

The impact of Penfield's claims varied in different contexts. It was greater in certain professional communities, particularly psychiatry and forensic work, than in academic research. But its greatest impact of all may have been on popular beliefs about memory, and through them on the creative arts. Penfield was perhaps the most prestigious and academically authoritative person to make claims for permanent and restorable memory traces, and his claims were more literal and specific than, say, the advertisements for home movie cameras. The idea that our experiences do not fade and die, but forever lie waiting to be reawakened, was practically intoxicating. And in fact, as much as Penfield found inspiration in the medium of film, so the medium of film found inspiration in him.

One of the weirdest and most intriguing projects inspired by Penfield's memory work was the avant-garde film corpus of Harry Smith. During the

1950s Smith was developing a variety of techniques of stop-motion anima-tion, particularly a "cutout" approach that used pictures from old maga-zines and catalogs. Smith stockpiled thousands of pictures, coding and cross-referencing them with extensive filing card systems. He drew on this enormous archive to create animation sequences, by taking a series of photo-graphs of scraps in various positions.

> I had everything filed in glassine envelopes; any kind of vegetable, any kind
> of animal, any kind of this that and the other thing, in all different sizes. Then
> file cards were made up. . . . This was all written on little slips of paper, the file
> cards—the possible combinations between this, that, and the other thing. The
> file cards were then rearranged in an effort to make a logical story. . . . Certain
> things would have to happen before others: Dog runs with watermelon has to
> occur after dog steals watermelon.
>
> I tried as much as possible to make the whole thing automatic, the produc-
> tion automatic rather than any kind of logical process.[33]

Smith's work procedure was in some ways evocative of how Penfield thought the brain created recordings: individual images were stored away, each representing a potential moment of a story, and each made from little scraps of records from real life—a cutout from a Sears Roebuck catalog, an advertisement torn from a magazine, a greeting card, a photograph. So perhaps it is not surprising that Smith knew and was fascinated by Penfield's work.

Smith's most famous film, *Heaven and Earth Magic*, originally released in 1957, was substantially inspired by Penfield's claims about memory and consciousness.[34] According to Smith's own description, the film depicts the heroine's toothache "consequent to the loss of a very valuable watermelon." She receives drugs from a dentist and is transported to "heaven," during which time her travels take her to "Montreal"—a reference to Penfield, whose *Epilepsy and the Functional Anatomy of the Human Brain* (1954; written with Herbert Jasper) fascinated Smith. The psychological effects of his sur-geries on conscious patients are clearly depicted here. A woman, who is often shown in profile and transparent, is injected with a giant needle; her head is opened and the effects of Penfield's surgeries are displayed—a sequence of images that seem intended to convey a stream of consciousness.

The raw materials from which Smith made his film—scraps from every-day life, including that most comprehensive and everyday compendium, a

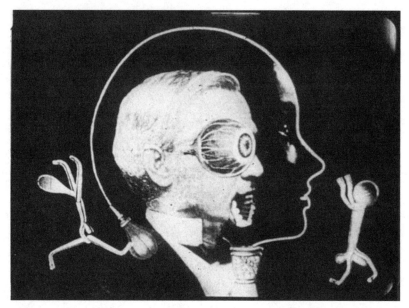

Still from Harry Smith, *Heaven and Earth Magic* (date not determined, c. 1957–62). Courtesy of the Harry Smith Archives.

Sears Roebuck catalog (from which Americans could buy anything from underwear to a free-standing home)—echoed Penfield's own claims about the scrappy nature of the records from the past that he called up to consciousness. The action is a dreamlike series of associated images with no continuous narrative flow or straightforward logic. Objects seem to flit by, placing viewers in the position of the hallucinating subject.[35] The structure of the film seems designed as an experiential lesson in the neurophysiology of fantasy and memory.[36]

Penfield inspired many such creative references, direct and indirect. Philip K. Dick's novel *Do Androids Dream of Electric Sheep?*—the basis for the 1982 film *Blade Runner*—even included a piece of technology named after Penfield. The novel, set in a postapocalyptic world in which mechanically engineered animals and androids comfort the diminishing community of "real" humans, poses the question of what defines a human being. In the story, humans use a personal device called a Penfield Mood Organ to choose and produce particular emotional states. The machine is effectively a household version of the technique Penfield practiced with his electrode. In the novel, however, these states are drawn not from the personal past of each individual, but from a mass-produced program for feeling particular emotions and longings—such as the "pleased acknowledgment of husband's superior wisdom in all matters," or "desire to watch television, no matter what's on it." The existence of a technology that mechanically triggers chosen emotions (emotions that in these cases are ironically those felt anyway by Americans in contemporary media representations) undermines the distinction, which people in the novel cling to, between "natural" humanness and the artificial character of machines. Dick's satire is one that Penfield would surely have appreciated, especially given his interest in making broad humanistic reflections on the implications of his technical work. The invention amounted to an ironic extension of Penfield's own intellectual expansiveness—of what he saw as his fundamental intellectual humanism.

Penfield's memory claims were also put to work in the theater. Constantin Stanislavski's "method" taught actors to mine their own sensory and emotional pasts in order to produce "real" modes of acting in the present. Stanislavski himself had drawn on Theodule Ribot's *Psychology of the Emotions* (1897), but Lee Strasberg, founder of the Actors Studio and an immensely influential force in acting in the postwar decades, updated the scientific underpinnings of the method. Strasberg described the actor's central challenge as that of producing "not just the words and movements he practiced in rehearsal, but the memory of emotion. He reaches this emotion through the memory of thought and sensation." Although the physiological nature of these actions was not well understood, Penfield's memory experiments, which he described, showed that these were indeed real past experiences revived in the present. Of course, Strasberg wrote, the triggers for what Penfield had termed the "experiential" revival of the past were different in daily life from the extraordinary conditions of his operating room. "In real life this process is stimulated by some conditioning factor. . . . For example, when someone tells you that he met a particular individual for whom you have strong feelings . . . your heart starts pounding. You will find yourself reacting merely to the mention or suggestion of that person, even in his absence." The job of the actor was to manage these conditioning factors so as to produce the appropriate feelings at will.[37]

Penfield's work can be seen as part of a broader shift in thinking about the personal past that was taking place in the 1950s. His memory claims proved peculiarly resonant as vanguard instances of that shift. They found their way into textbooks and documentaries about the mind in the 1960s and into the 1970s, including his own widely read autobiographical and philosophical

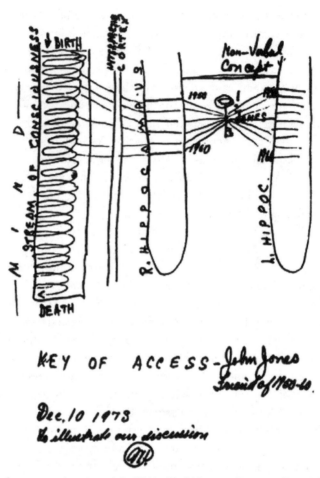

Sketch of memory mechanism made by Wilder Penfield in 1973. Courtesy of Brenda Milner.

reflections, written shortly before his death.[38] Memory continued to be prominent in his thinking and writing even in the 1970s. In 1973, after a discussion with Milner about antegrade and retrograde amnesia, he drew a sketch representing the relation between different parts of the brain in healthy memory storage and retrieval. At the left of the sketch is the (theorized) neural record of consciousness, containing moment-by-moment records of a lifetime of experience. In the middle and to the right of the sketch are the two hippocampi. As Milner wrote that Penfield told her, "If you are trying, for example, to remember something about John Jones, who was your friend between 1950 and 1960, then in some way, via the interpretive cortex of the temporal lobes, the hippocampi give you "keys of access" to those past recorded experiences."

Milner, who wrote about and published the sketch in 1977, used it primarily to demonstrate how engaged Penfield was in these questions up to the end of his life—she acknowledged that the formulation "will seem to most psychologists little more than a picturesque metaphor."[39] Earlier in the article she had noted, in what sounded like an anticipatory response to Penfield skeptics, that the memory effects he had produced in the operating room had been rare in other surgeons' experience, but this was not surprising once one considered that Penfield alone had such a "tremendous clinical series of patients undergoing temporal lobectomy," and few people applied themselves "with such unflagging scientific interest to document carefully the subjective reports of their patients in the operating room" instead of dismissing them as epiphenomenal.

Indeed, the plausibility of Penfield's memory work was becoming tarnished in the late 1970s, because it seemed to stand in the way of a very different approach to mind and memory being promoted by the new field of cognitive psychology: autobiographical memories were to be seen not as inert impressions but as dynamic, constantly changing entities. Academic psychologists had never embraced Penfield's claims, but cognitive psychologist Elizabeth Loftus, one of the pioneers of late twentieth-century forensic studies of memory, the central pillar of which was the claim that memories were fundamentally reconstructive and unreliable, discussed Penfield's work in her own writings. She pointed out that the kind of experiences that had generated the extraordinary attention for Penfield's claims had in fact occurred in very few of his patients—some 3.5 percent, by her count. (This percentage is much lower than Penfield's own report of 15 percent; the difference may be that Penfield referred to a subset of his patients whom he called "hippocampal," that is, those with lesions in particular locations, who might in principle have experienced returning memories by having their "memory cortex" stimulated.) Loftus's point was that the responses on which Penfield staked great claims about memory were in fact very rare, and quite possibly unrepresentative of normal processes. Moreover, Penfield had offered no evidence that the materials were in fact memories rather than hallucinations of some other kind. Penfield, she argued in 1980, had found nothing (beyond his own interpretative tastes) to persuade him or anyone else to move from his initial interpretation—that his patients were experiencing memory-like hallucinations based on various kinds of psychic material—to his later one, that what they were reporting were memory records, much less memory records of perfect fidelity.[40]

However, in midcentury the best-known experiments that sought to bring the past back to life were not Penfield's, famous as they were. They were made by a previously obscure businessman and a housewife in a little town in Colorado. When their experiments were publicized, the most intimate details of the housewife's past life—or perhaps one should say, her past *lives*—became the talk of the nation.

5

THE THREE LIVES OF
BRIDEY MURPHY

On November 29, 1952, an amateur hypnotist began an experiment with a friend. It began straightforwardly enough, as she tried to focus her attention "back, back" to revisit earlier times and other places. But then the experiment took an unusual turn. The hypnotist, a Colorado businessman named Morey Bernstein, suggested to his subject, a housewife named Virginia "Ginni" Tighe, that she might progress still further back, beyond childhood and even her own birth, to find herself in a former life. There was a period of silence. Then Tighe began to speak in an Irish brogue. She introduced herself by another name and started to describe her life—in early nineteenth-century Cork.

This was the beginning of an extraordinary account of the scientifically managed remembering of a "past life," and it burst onto the public stage a few years later to throw American society into a veritable "hypnotizzy." Bernstein's book describing the sessions with Tighe topped the *New York Times* best-seller lists, and copycat experiments swept the country as everyone debated whether the experiments were valid. Researchers were dispatched to Ireland to check the plausibility of specific points in Tighe's story; psychologists were recruited to evaluate the records of the experiments; and journalists tramped through Chicago and a small town in Wisconsin, where Tighe had grown up, to study her own past.

The Bridey Murphy affair partook of the popular revival of strong convictions about the powers of memory. It advanced the idea that previous phases of life were perfectly archived somewhere in the mind, brain, or soul.

It was apparently possible to return to previous eras and to re-experience them without the filter of hindsight and the fading of memory. One could temporarily *become* one's past self.

At the same time, the controversies opened a window onto what history as well as memory could mean outside the preserves of academic disputes. The understanding of historical knowledge introduced in the Bridey controversies was antithetical to the assumptions about historical evidence promoted by academic traditions. It emancipated everyman from the archive by making every individual into an archive of world history. Clinical psychologists, journalists, local historians, and genealogists all weighed in, sometimes even doing their own research according to their own lights. Indeed, the controversy revealed attitudes toward memory and history that would otherwise be hard to document because they were generally not written down or otherwise publicly articulated.

New Lives

Bernstein came from a family of entrepreneurial optimists who, as his obituary put it, had always "taken a lot on faith."[1] His grandfather emigrated from Russia, then moved west to found a scrap-metal operation in Colorado. Bernstein inherited the business and made it into a metal construction company. The "reincarnation," as one news article wrote, of scrap taken to steel mills from automobile "graveyards" was one of the fastest growing areas of industrial-commercial growth in the decade after 1945, and Bernstein was on the crest of the wave.[2]

Bernstein witnessed hypnosis at a party in the early 1940s and was intrigued. He read everything he could find about it and practiced on his wife (she said it relieved her headaches). Then he tried it on his friends, recording his experiments on reel-to-reel tape. He was particularly interested in what he read about hypnotic age regression, which seemed to him to be "one of the most fascinating phenomena in the field." According to most hypnotists, there were two kinds of regression, one more controversial than the other. Hypnotic subjects might recall a particular experience from some remove, as though they were bystanders. But the experience was even more striking in "total" regression, where the subjects seemed actually to be *reliving* the past experience, as with the traumatized World War II victims whose experiences everyone was hearing about at this time. Bernstein was interested in this second kind.

Bernstein's readings soon took him further afield, down the exotic paths that hypnosis could explore in late 1940s America. The topic of reincarnation caught his eye. It was in the news because of China's re-annexation of Tibet in 1950, and more generally the spread of communism in Asia. Eastern religions were frequently portrayed as a cultural wall whose reinforcement could slow the spread of communism in Asia and the Indian subcontinent. Such articles unfailingly mentioned the young Dalai Lama, believed by Tibetans to be the current incarnation of previous Dalai Lamas.

Bernstein himself, however, was drawn to reincarnation not by Eastern philosophy, but by its role in an eclectic assortment of American parapsychological writings.[3] These led him to Edgar Cayce, a self-professed psychic who had recounted thousands of clients' past lives. His thousands of consultations, and the several books about his work, supplied some of the nascent ingredients of what we now call "New Age" beliefs.[4] Bernstein, initially outraged that such excesses would besmirch "sane" hypnosis, pursued the Cayce story. Cayce was dead, but his son maintained a vast archive (which still exists).[5] After interviewing several of Cayce's clients, Bernstein grew less skeptical. He was not yet convinced, but he did want to try his own experiments.

Bernstein met Virginia Tighe after a dance one night, when friends invited him back to their house for drinks. Bernstein was asked, as usual, to perform hypnosis, and Tighe, he later wrote, "[went] completely under."[6] He was amazed by her extraordinary hypnotic potential. Some weeks later, when the Bernsteins invited the Tighes over for dinner, they tried another session. Tighe again quickly fell into a deep trance. Impressed once more by her responsiveness, he asked if she would be open to being regressed to a point before her birth. She agreed.

Once Tighe was clearly entranced, Bernstein asked her to go "back, back, back" to childhood. Then he suggested she go further, to "another time." After a period of silence, she announced rather abruptly that she had scratched the paint off her bed. Then she remarked that her father was a "barrister" in town. Bernstein had not been particularly impressed by the first utterance, assuming it a memory from her own childhood. But Tighe was neither well educated nor well traveled, so he was surprised to hear the word "barrister." Then, Bernstein later reported, a "subtle brogue" crept into her voice.

Tighe's speaker soon gave her name as Bridey Murphy. She detailed her life, her death, and even her experiences afterward. Apparently she had waited for her father to die so that she could point out to him that she hadn't gone to purgatory. When asked about the spirit world, she complained that "you couldn't talk to anyone for very long" because "they'd go away."

Bernstein played the tapes for his friends. They were all convinced he was dealing with something extraordinary.[7] Over the course of five more sessions, Bridey reinforced the impression. She related that she had been born in 1798 to a Protestant family in Cork, the daughter of barrister Duncan Murphy and his wife Kathleen.[8] She married and moved to Belfast, where she died from a fall in 1858. The names, dates, places, and events were striking in both their breadth and their specificity of detail.

Bernstein decided to write a book about the experiments. As he worked through his notes, he sent Tighe a series of questions aimed at confirming that she had had no ordinary memories that could account for her hypnotic statements. He also asked about her reading habits. She allowed that information might be "secreted away" in the back of her mind, but she told him she knew of no textual or autobiographical source that could explain them. Yet Tighe began to be nervous about drawing the wrong kind of fame: her letters regularly reminded him to protect her anonymity, because her relatives would "not approve." She would reconsider only if her past life became accepted as proven fact.[9]

In the summer of 1954, *Denver Post* reporter William J. Barker heard of the experiments. He approached the pair and reported on their experiences in two articles for his paper. His report described Bernstein as a "cool, conservative fellow," a "good-looking, hard-working businessman on the board of directors of three firms and a bank."[10] Barker portrayed Bernstein and Tighe (whose identity he sought to protect by using the pseudonym Ruth Simmons) as skeptics who had resisted the strange revelations until forced by experience to take them seriously. He included several striking photographs, including one of the entranced Tighe sketching the Antrim coastline of Bridey's home.

Soon after the articles were published, Tighe wrote to Bernstein that word had gotten around Pueblo that she was the hypnotic subject. "Many, many people have called," she told him, "and asked about them & how I felt."[11] The fascination proved to be broader than just her neighbors. There was such a large response to the *Post* article that Barker published another. News of Bernstein's book in progress spread quickly and triggered thousands of advance orders—a full year before the publication date. He already had a contract with a small publisher, but Doubleday soon bought the rights and made preparations for a best seller.[12]

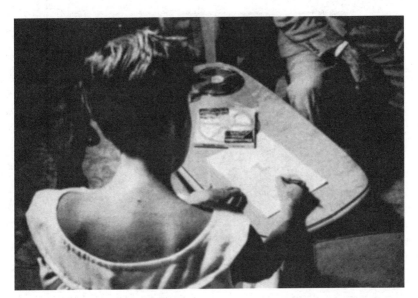

Virginia Tighe, entranced and acting as "Bridey Murphy," sketching the outline of the Antrim coast of Ireland. *Denver Post*, September 1954.

As the book moved through publication, Bernstein financed the production of a twelve-inch long-playing record with excerpts of the first hypnotic session.[13] He confessed privately that sales of the LP were dismal,[14] but it did draw some critical attention. One reviewer was struck by the contrast between Bernstein's "suave, insistent questioning in the authoritative tone of a modern business executive" and "Bridey's gasping, agonized, wheezing replies." There was high "drama" in the possibility that "a young girl from a small Western mining town [could] recall past happiness" in nineteenth-century Belfast.[15]

Bernstein's book appeared the following month, featuring a picture of him with his now famous tape recorder on the back of the jacket. The first 10,000 copies were snapped up, and Doubleday printed 200,000 more. The book stayed on the *New York Times* best-seller list for months, topping the nonfiction list. It would eventually appear in thirty languages and thirty-four countries.[16] Work began on a feature film, in which a series of actresses would play Bridey at various stages of her life and well-known actress Teresa Wright would play Virginia Tighe.[17]

In the midst of all this, a giddy salon culture developed. Hypnotists staged sessions in homes, halls, and theaters, selling "existences" for twenty-five

Hypnotist carries out "Bridey Murphy" experiments on members of Kappa Alpha Psi fraternity in 1956. Courtesy of the *Chicago Defender.*

dollars apiece; parties were centered on home experiments, and students staged events at fraternities.[18] There was a humorous side to this nationwide preoccupation: party invitations instructed guests to "come as you were," and cartoons depicted new parents welcoming newborns with the salutation, "Welcome back!"[19] Liberace took the opportunity to stage a flamboyant Bridey performance involving many costume changes. But all of this belied the seriousness with which eager "soul searchers" strained for "glimpses" of former lives.[20] There were darker reports, too: a teenager reportedly killed himself, leaving a note saying he wanted to explore reincarnation "in person."

Meanwhile the reviews flowed in. Some were bemused, but others saw the book as having a broader metaphysical significance. William Buckley, writing anonymously in the first issue of his *National Review*, pronounced Bridey plausible and declared her story evidence of the "philosophical failure" of twentieth-century intellectual life. America had suffered a long "Age of Materialism, whose doctrine has offered neither meat nor drink for the human spirit," and Bernstein's experiments were helping to prove that "the universe is . . . more complicated than Epicurus, De La Mettrie or Marx supposed." Bernstein himself had not voiced an opinion on whether Tighe was

"Come as you were" party, ca. 1956. Photographer: Carl Iwasaki; Time and Life Images. Courtesy of Getty Images.

truly Bridey Murphy; nor had he explored other possibilities that occurred to Buckley—that Tighe might be a "medium for another spirit," or that Bridey might be the result of an "unrevealed" but wholly natural "capacity of the human mind."[21] Nevertheless, the tapes were "astounding," particularly since many (though not all, Buckley carefully noted) of the historical facts had been "confirmed." Not all defenders of spirituality were so open-minded. In a regional Catholic journal, one Reverend Henry Atwell spat that Bridey "contradicts such doctrines as original sin, particular judgment, the fact of purgatory, the eternity of heaven and hell, and the resurrection of the body."[22]

From an Irish American perspective, the Murphy tale took on a different appearance. It seemed to reinforce a naive and false, if romantic, history of Ireland all too familiar in the American context. Irish American writer and essayist Walter Ready thought the story indicated the cultural impoverishment of the public, who would have dismissed it had they been educated in anything beyond the crudest stereotypes. It brought "news of an Ireland that never was, save in the minds of the uninformed and the vulgar." Although he was primarily concerned with the substance of the book, its style alone was offensive enough to warrant Ready's denunciation: from it there "emanated" an "effluvium" of "tastelessness and dullness" and an unpleasant "air of do-good earnestness." He walked the reader through a few passages, those describing Bridey's courtship, her house, and a wake she attended. They were, he wrote, no less than insulting. For instance, Bridey described the wake as an event at which everyone "drinks tea and is unhappy." Such a statement, Ready declared, betrayed no familiarity with Irish life except "in the most general way, the way that everybody knows two Irish words, like banshee and colleen." He took the trouble to correct Tighe by recounting Irish grieving practices in some detail: how the body of the dead person was laid out in a room, where candles were placed, and the visits from friends and relatives.

On this reading, the whole episode was a discouraging sign of the state of modern American society. Far from being an authentic voice from beyond the grave, Bridey Murphy was a crude set of stereotyped assertions that shouted down subtler, truer understandings of Irish history. There could—there *should*—have been a "Bridey Murphy" echoing out of early nineteenth-century Ireland: a rich memory cherished and retold by later generations. Bernstein could have told of bygone times that lived on in common memory, like the English John Bull—a sort of "everyperson" of the Old Country. But Bridey destroyed the possibility of such a collective memory by stamping anew in people's minds the most ignorant stereotypes. "The shame of it," Ready concluded bitterly, "is that yet more harm will be done to the Irish memory by this deal."[23]

Given this reception, it is worth noting that the history Bridey narrated was *not* obvious in certain respects. It was surprising that she was an Irish Protestant, because Americans tended to equate Catholics with Irishness. Nor was Bridey chafing under British rule or struggling to make ends meet. Even stranger, her father was a barrister—a virtuoso of what was essentially an imposed English legal system. The idea of a recovered persona was typically bound up with the notion of a repressed, perhaps even *suppressed* authentic self, but as a Protestant daughter of a barrister Murphy was more aligned with the imperial oppressor than with the "true" ethnicity denoted "Irish." It is also suggestive that Murphy was born in 1798, the year of the biggest Irish rising between the seventeenth century and the twentieth: that event had led to the Act of Union, ending dreams of Irish emancipation for generations by joining Ireland and Britain into a single realm.

Overall, the Bridey "spell" proved powerful. As *Life* magazine testified, its influence was rapidly stretching "deeper and wider as fast as the written word, awed gossip and the televised image could spread it." Bernstein's book was selling faster than ever, and books on parapsychology saw their sales rise too. Melvin Powers of California, a distributor of books on hypnotism, found his business multiplied "twenty-five times over." Indeed, it is possible that the Bridey Murphy episode played a role in making the paranormal a publishing sensation into the 1960s and 1970s. According to another bookseller, Bridey had become the "hottest thing" since Norman Vincent Peale's *Power of Positive Thinking*.

The Letters

Bernstein was soon deluged with fan mail, ranging from a few lines of appreciation to substantial letters. For example, Perry Jester, a retired diplomat, wrote to say that millions of "soul-hungry" Americans were "searching for a spiritual interpretation of life that can be made acceptable to their minds—overlaid as they are with heavy veneers of scientific curiosity and materialism"—and traditional religion was unsatisfying.[24] Bernstein had confirmed for him that there was no "'iron curtain' lowered between this world and the next." One had only to exert oneself to find the "keys" to "communication" with the next world.

Jester's desire for a substitute for orthodox religion was shared by many other correspondents. One writer from Texas sent a skeptical news clipping and a cover note urging, "Don't let them stop you. When this is proven as Truth, it will ruin some of the religious systems and they are going to fight this."[25] Tighe, too, had received a letter from a Pueblo steelworker who reflected that "one cannot blame the 'clergy' for their interest. Bernstein has struck at the very foundation of their racket."[26]

Some of Bernstein's thousands of correspondents had their own theories to contribute.[27] One of these claimed that the mind had "three parts," and in the trance the subject passed through all three of them. The first, she wrote, "is conscious mind, [the] second sub-conscious, and the third which isn't mind at all really; the entity itself or undying spirit." Another correspondent reported how excited her reading group was by Bernstein's book; their interest lay in heterodox approaches to "prayer in all its aspects."[28] Subscribers to the Theosophical movement saw the experiments as an opportunity to go mainstream, and their journal *Ancient Wisdom* began a lengthy series on Bridey Murphy. Morey Bernstein himself corresponded frequently with the editor, Charles Luntz, and wrote a preface to Luntz's book *The Challenge of Reincarnation*.

Many readers wrote to ask for more. Students wrote in from several universities to suggest or request research publications. One of them, Charles Tart, then an electrical engineering major at MIT, eventually became a prominent figure on the New Age sciences scene.[29] A bank president reported that he knew a woman who could do regression through many successive lifetimes, and he wanted suggestions for what to read to learn more.[30] Others sent practical suggestions for other experiments or new techniques. For instance, one letter suggested that Bernstein try ether—Bernstein had tried "narcosynthesis" without success, but ether required no IV. Someone with the serendipitous name of Phyllis Krystal wrote to say that she had visited Cayce and described her own "reverie" technique. Another Cayce affiliate wrote to describe experiments in which he had taken part, reviving a previous existence in Africa about AD 350 and an ability to speak the extinct language of that time.[31]

When one considers that Bernstein came to his readers' attention as a powerful restorer of personal memory, it is interesting to see how much of their own lives people entrusted to him. Some letters were several pages long, as people poured out autobiographies medical and personal.[32] One woman asked Bernstein to help her access memories of a previous life that might have been better than the one she was living, because this one was so unbearable. Many others hoped to interest Bernstein in hypnotizing them, in search of their former lives or to access long-buried memories from their childhoods. Most offered a kind of curriculum vitae of their qualifications as extraordinary rememberers. One bragged that he had "always been possessed of a sense of the past 'so strong it could be rolled out flat and cut up into awnings.'"[33] Another volunteered her services as a regression subject, explaining that her unusually good memory (and her husband's permission) qualified her as a candidate. These correspondents had widely divergent theories of memory to offer. One letter asked if hypnosis could be used to reestablish "physical memory pathways," helping to "restore the muscle memory where it has been destroyed (or shocked) by polio."[34] Another put forward an elaborate theory of multiple life "cycles" and techniques for what one might call (albeit anachronistically) translifetime "personal growth."

As one might expect, in these letters there were many stories of successful reincarnation experiments. Bernstein replied to them all, with suggestions and sometimes doubts. One man wrote from Ontario to describe experiments on his daughter that led to memories of a past life in a nearby town. The memories seemed too mature for her age of fourteen. Bernstein worried that although names and places seemed to check out, that the memories related to a place near where the family lived meant she might have picked up the information some other way without realizing it; the father dismissed this possibility.[35] A husband and wife wrote together to report that they had successfully accessed their past lives through a technique they called "retrocognition," which was done in a waking state, though the information had to be written down quickly before it faded from consciousness.[36] Another correspondent described how memories of her past returned as she was "being 'processed'" on an electropsychometer by a "Scientology auditor," revealing a past life as a minor ruler in Egypt before the time of the great pyramids. *Life* published photographs of several such experiments—a middle-aged secretary curled up on the floor like a baby, a grown man miming milking a cow (a chore his grandfather carried out daily as a child), and a "co-ed" at UCLA performing hypnotic activity before a rapt audience.

Some correspondents were motivated by concerns with which the professional historian may feel a sneaking sympathy. Several suggested that past-life regression could help people learn more about the historical past—not just the past lives of individual subjects. One R. Bowerman wrote to say that his father and a photographer had been the first to find and open the Stone Age mounds along the Calapooyia River near Albany, Oregon. The artifacts of these explorations had been given to the University of Oregon, though the university had not been told how the discoveries were made. The truth was that "this was accomplished through the revelations of a hypnotic subject, who revealed a giant race of men, a pygmie group and a medium highth [*sic*] race who inhabited this Country many years before the so-called Indians arrived here."[37] Irene Smith Johnson wrote to Bernstein because she was "compiling material for some historical books and there are some facts which it seems I can not prove." Since one of the books was about her own family, back in the pioneer days, she wondered if age regression could reveal facts about her family life generations back.[38] More staidly, professional historians and genealogists offered to help find conventional kinds of evidence of Bridey's existence, confident that they were better suited to the task than civic record keepers—and of course journalists.[39]

It seems that the Bridey experiments did not introduce extraordinary new ideas about the past into public culture. Rather, they created an opportunity for many people who already held similar beliefs—and perhaps had even carried out similar experiments—to express their views. Bernstein was not someone whose unusual claims about the nature of memory, history, and identity challenged the conformity of staid 1950s America. It is more appropriate to see him as starting a public conversation about beliefs and practices that were already thriving in private. Many deeply rooted understandings of memory—of the reality of past lives, and theories of what it was possible for an individual to know—were already available in the broad population to support the Bridey assertions, but they had been largely invisible. When they became public in the late 1950s and 1960s, it seemed as if they were coming into existence for the first time.

The Investigations

As Bernstein's book climbed the best-seller lists, investigative journalists flew to Ireland to check Bridey Murphy's story against the historical record. Among them was Bill Barker, who had written the first Murphy report.[40] Barker's new report represented the journey as a trip back in time: the seats of the train reminded him of Hitchcock films from years past, the town of Limerick looked like a Dickens setting rendered in color.

What Barker and the other journalists brought home was a confusing mixture of the unverifiable, the incorrect, and the confirmatory. The only clear result was an intensification of the debate. *Life* published a lengthy article in mid-March, including a huge multipage spread summarizing the reporters' findings.[41] The magazine had long been a skeptical voice in the episode, but not a scoffing one. The reason for the excitement, *Life* suggested, lay in the plausibility of the experiment and the experimenters. Bernstein it described as a businessman of "impeccable reputation," certainly "no brooding Svengali." As for Ginni Tighe, she was an intelligent, trustworthy young woman with no mental health problems—indeed, residents of Pueblo, all of whom knew her real name, so respected her that they steadfastly protected her anonymity. But the most persuasive element was the book itself. The very amateurishness of the prose was compelling, *Life* concluded: Bernstein's "frankly awed, what-have-I-wrought style" was "far more convincing than any professional glibness." The other factor was the wealth of "subtle and plausible" Irish detail threaded through Bridey's recitations. It was hard to imagine a midwestern woman who had never visited Ireland (and of course had not studied the Ireland of a hundred years before) producing such information. Although not everyone would be convinced by the book, "there are few who can go through its 256 pages without being shaken or impressed." Now the magazine went through all those pages itself, determined to score the claims they made.

Life tabulated Bridey's statements as claims of fact, comparing them point by point with what Barker, Ernie Hill (of the *Chicago Daily News*) and Ruth Lynam (of *Life* itself) had turned up. For instance, Bridey Murphy claimed to have lived in a wooden house called the Meadows, but in the early nineteenth century the scarcity of trees meant that most houses were made of stone. Yet wooden houses were not unheard of. She referred to having a painted iron bed in 1802, but *Life* claimed there were no iron bedsteads in Ireland before 1850. She pronounced some words in a way that, skeptics claimed, no native would have done. And she referred to a church, St. Theresa's, which did exist—but only since 1905. The very intensity of the controversy allowed for a detailed accounting of such ways that reporters' research diverged from her story, or at least failed to verify her claims.

The summary of these findings used the vocabulary of fact, but one might argue that it was unlikely anyone could be *certain* there were no iron bedsteads in Ireland before 1850, and that no church by the name of St. Theresa's existed then. Indeed, Bridey supporters did research each rebuttal claim. They produced evidence that iron beds did indeed exist in early nineteenth-century Ireland. But by the time this research was published, about 1960, the national media had moved on.

Moreover, despite *Life*'s own skepticism, the table itself did not consistently contradict the Bridey narrative. It confirmed several of her statements, and sometimes on points of information that were culturally, geographically, or historically obscure. Bridey often seemed eerily accurate. Her sketches of the Antrim coastline were perfect. So were the descriptions of what one would see on a journey between Cork and Belfast. Her references to various businesses, too, checked with ones that had indeed existed: a grocer named Farr, a rope-making company. Had the Irish research consistently contradicted Bridey/Tighe, this might have been the end of the matter. But these mixed results seemed inconclusive.

Life now looked to psychiatrists and psychologists in the hope that they could resolve the puzzle of the apparent memories. Two of these experts, Jerome Schneck (past president of the Society for Clinical and Experimental Hypnosis) and Lewis Wolberg (director of New York's Postgraduate Center for Psychotherapy), part of a larger panel of psychologists discussing the issue, were quoted at length. Both explained that hypnotized subjects drew on all the resources within their "subconscious memory." As a result, Schneck told *Life*, under regression someone might be able to speak a language he had not heard since childhood, or quote verbatim from a book he had not seen since age three.[42] But even if these performances came from real memories, that did not mean that they were true to the original experiences. Hypnotic subjects often used memories to "fabricate" what they said and did in a trance. This was especially true of "somnambules," people who were especially susceptible to deep hypnosis. And there were even a few people who were so suggestible that "the most experienced and scientific hypnotist cannot tell immediately whether the result . . . is feigned or not."[43]

This pointed to a possible new approach to evaluating the Bridey story: a historical study of Tighe herself. Schneck and Wolberg told *Life* that the expensive project of sending reporters to Ireland to check Bridey's historical plausibility was "roundabout." The more "direct" way was simply to go into Tighe's own "childhood and development," as well as "her parents and [her] relationship with them." As John Dollard, a Yale psychologist with an interest in both social psychology and psychotherapy, put it, a psychological study of Tighe "would take all the mystery out of this case." The indications of previous existence would fall away, revealing a subtle dynamic of wishful thinking produced by the "remarkable power of unconscious mental life," powerful enough to "bilk not only a naïve hypnotist, but also the subject."

The several fact-finding projects launched in the wake of Bernstein's book therefore relied on different, even conflicting, assumptions. The rationale behind an Irish trip was to find facts that confirmed or conflicted with those claims that were open to confirmation or falsification—names, places, marriages, and buildings. If the issue was one of historical falsification, the place to go was Ireland. Only there could one find that the landmarks she had described did not exist; that her material surroundings (house, bed) could not have been what she said they were; or, if detailed birth records had been available for the eighteenth century, that Bridey herself had never been born.

For the psychologists, on the other hand, biography was the best route to the truth. They suspected that the claims arose from Tighe's (not Bridey's) memory. The way to confirm this suspicion was explore Tighe's childhood, not Bridey's. If sources for her statements could be found in twentieth-century America, *their* claims would grow more plausible. Better yet would be to find a basis for the story specifically in Tighe's childhood—her own experiences, family, and friends. If elements of Bridey's life were found in Tighe's, it would be a clear sign that Bridey was a somnambulic "fabrication." Nothing from her childhood could conclusively falsify the Bridey Murphy story, but it would provide a competing, naturalistic explanation of greater credibility.

Where would such information come from? Ray Brunner, a newspaper reporter in Toledo, Ohio, searched old movies and novels looking for a Bridey prototype. But the psychologists discounted the possibility that Bridey came fully formed out of a single fictional source. They wanted access to Tighe's mind because they thought that Bridey was the product of many bits and pieces of memory, knitted together into a single narrative. They proposed doing extensive interviews with Tighe, and—the irony of this does not seem to have occurred to them—conducting age regression experiments. The same psychologists had chided Bernstein for being too naive to realize that all information procured under hypnosis was a fabrication. Yet they were willing to claim that the raw materials of that fabrication could show themselves under the same techniques—presumably, because they were more skilled in applying the technique. They were sure that if Tighe "could completely reveal her early life to them, preferably under hypnosis," this would lead to a scientifically legitimate resolution. But both approaches, needless to say, required Tighe's consent. And she refused.

There was, however, another way to develop a biographical account of Tighe: by applying to her biography the techniques of investigative journalism that *Life*, the *Post*, and the *Daily News* had used in Ireland. In the absence of the subject herself, the *Daily News* performed an analysis of the texts surrounding her life. At this point the *Chicago American*—a Hearst newspaper with a reputation for sensational journalism in defense of the common man—took up the cause. Reporters fanned out through Chicago and Wisconsin to the neighborhoods of Tighe's childhood, poring over documents and interviewing old friends and members of her extended family.

Tighe was distressed by this development. She had been concerned from the start that some relatives in Chicago never learn of her involvement in Bernstein's experiments. They would be "hurt" by the news, she feared, because the idea of reincarnation would offend their Christian sensibilities. Until late spring, all newspapers had referred to her by the pseudonym Ruth Simmons. But the intense scrutiny of the case in the Chicago newspapers did indeed bring her involvement to the attention of her family, who recognized her and gave interviews to journalists. Her sister and her aunt telephoned Bernstein's publisher, and then Tighe herself, castigating her for what they saw as irreligious, self-aggrandizing behavior. The intense and often mocking scrutiny was hard to bear. In late May Bernstein wrote to say, "You were a small boat on a very rough sea, but you navigated deftly. And now the weather is clearing."[44]

The weather was not clear for very long. On May 27 *Chicago American* headlines screamed, "Was Life Here Her Ireland?" It turned out that some of Bridey's experiences were "remarkably like what happened in Chicago to the girl who now lives in Pueblo Colorado." A series of articles delivered a great deal of fine-textured information about Tighe's past that seemed strikingly similar to details of Bridey Murphy's story. The research was carried out not by Chicago reporters alone but by reporters in collaboration with an unlikely investigator: Wally White, pastor of the Chicago Gospel Tabernacle, the church Tighe had belonged to as a child. At that time it had been a rapidly expanding evangelical church. White worked to make it a center for evangelical activity—not only sermons, but film screenings, educational activities, and missionary work.[45] In a series of articles in the *Chicago American*, White now portrayed himself as having a responsibility to protect the church from the dangers posed by the Bridey Murphy story. He was not the first to make a link between past life regression and loss of faith: Bernstein himself, in his initial interview with the *Denver Post*, had described his own loss of faith as the context for his readiness to embrace the stranger possibilities of memory and hypnosis, remarking that the more he had "looked at the world around me," the more he had seen religion as a "mythology."[46] But White decided to pursue the case because he learned that Tighe had once been a member of his church. If the Bridey story was antithetical to Christian teaching about death, then he saw it as his responsibility to investigate.[47] Support for the "old theory of reincarnation" represented an attack on "one of our civilization's foundation stones—faith in the . . . scriptures." White was particularly concerned about the epidemic of copycat regression experiments. Such experiments were dangerous not only to the mental health of subjects, but even to that of *potential* subjects, as witnessed by the case of the young man who had credited Bridey with his decision to commit suicide.[48]

White knew many members of Tighe's extended family from his church. Now he asked them about incidents from her childhood that might have their counterparts in the Bridey narrative. Tighe had a sister in Chicago named Marie (pseudonym Helen in Bernstein's book), and she discussed her own childhood memories with White. She claimed, for instance, that Tighe herself had been spanked for scratching the paint off her metal bed—just like Bridey.[49] White and several *Chicago American* reporters scoured all sources of information on Tighe's childhood to find other information of this kind. They interviewed relatives, friends, and classmates, studied city directories, and examined the books they were told Tighe had read. They were able to find marked similarities between incidents and names in Tighe's and Murphy's lives. Although they had not been able to find every scrap of material relevant to the early life of Ginni Tighe, what they did find supplied the ingredients for a naturalistic explanation of her experiences. Not wishing to discredit Tighe herself, White took care to report approvingly that she was as "troubled" by the phenomena as everyone else—just as the "western world" and its "traditional religious beliefs" were troubled. He assured readers that he had spoken with her and that she had rejected the possibility of reincarnation.

May and June saw a stream of further revelations, more telling in their detail than in any one damning disclosure. Bridey had told of having an uncle with the peculiar nickname of Plazz; so, it seemed, had Tighe.[50] Tighe claimed she had been brought up by parents of mixed European and Eastern European background, but these turned out to be her foster parents; her biological parents, who raised her until she was four, were Irish American. She was said to have had no trace of an Irish accent, but her old school friends claimed she was known for being able to produce an authentic-sounding brogue. The *Chicago American* had also consulted Irish speakers who claimed that Bridey pronounced certain common Irish words in a strikingly American way.[51] It also turned out that her aunt was Irish, and that shortly before the first experiment in November 1952, she had died and left Tighe $2,500 in her will. The implication was that the legacy of her aunt's life (literally and figuratively) was in her mind in the autumn and winter.[52]

The Chicago sleuths also investigated Tighe's childhood neighborhoods. They found a store near her home that had a name similar to one Bridey mentioned and that sold similar merchandise. Bernstein had sought verifiable historical references from Bridey in an effort to produce a set of information that could be measured against the historical record. One reference he obtained was to a store called Cadenn's House, where Bridey had bought lingerie. The *Chicago American* reporters, after combing several old city directories, found (in the directory for 1928–29) that a Kaden dry goods store was in operation from 1925 to 1940 near Tighe's home. She would have had to walk past the store to go to the cinema. They even tracked down the owner, Bernard Kaden, who confirmed that he had often put women's clothing in the display window. Even the name "Bridey Murphy" could be found in her old haunts—this was the maiden name of a woman who lived across the street from Tighe's childhood home.[53] Finally, on June 7 the *Chicago American* published a point-by-point list of these findings alongside the claims they rebutted. Its point was clear: Bridey Murphy was the result of a "mental patchwork," cobbled together from memory scraps by Tighe's own "subconscious mind."[54]

Psychiatrists who had previously longed for access to Tighe's subconscious memory were thrilled by this new information. Lewis Wolberg, the expert hired by *Life*, applauded the "pioneer detective work" and pronounced it an adequate substitute for a direct examination of Tighe's mind. Chicago psychiatrist Alfred Solomon described the revelations as a "far more intelligent" way of checking the story than sending reporters to Ireland.[55] Psychologist Milton Kline published a whole book of "researches" based on the reports.[56] And Milton Erickson, an eclectic psychiatrist who often used age regression, thought the reports were an instructive example of how regressed subjects used "all past events and experiences along with the thoughts and ideas encountered by contacts with others." Erickson called Bernstein, with pointed irony, a "competent dealer in farm machinery."[57] Not having any expertise in hypnosis, Bernstein had not realized that when he asked Tighe to go "back, back, back," he was asking her not to reproduce original records of her past experience, but to mentally flip through all the books she had ever read. It was "simply," he said, an "invitation to plagiarism."

Meanwhile, other regression subjects were drawing new kinds of conclusions from their own experiments. One Roberta Westwood of Chicago gave an interview to a *Chicago American* reporter about experiments with a local hypnotist. She was regressed to her childhood, then to a former life as a nineteenth-century teacher. But she saw her entranced state as one of "super-concentration" and "intense desire" to "please the hypnotist." Unlike Tighe, who had not remembered what she said during the experiments, Westwood did: "What I said was what I thought would please my hypnotist and the people watching."[58]

Tighe and Bernstein were appalled by it all. In late May or early June, he told her of strategizing sessions to decide on a response. "Lawyer Rubin and I huddled," he wrote: "The American articles were nonsense . . . [and] should [not] be dignified by threats or suits." She should "try to relax." Bernstein asked Tighe to give interviews as part of a coordinated response to these and the *Life* articles.[59] But Tighe drew the line. "No more personal publicity," she insisted. She could not bear the constant harassment by clergymen "who feel I have done a gross injustice to the human race."

Inevitably, however, in late 1956 and early 1957 the news coverage finally began to wane so that Tighe and Bernstein felt less under siege. But Tighe's family continued fighting about the episode and its publicity. She never reconciled with the Chicago branch of her family, and their feud continued well into the 1970s. Tighe herself later recalled an argument with one Chicago aunt, explaining how the aunt's "fanatical viewpoint" about religion had led her to exasperatingly dogmatic positions about memory. "Why do people insist on believing what someone else tells them?" she exclaimed. "Why can't they see that the person who told them (parent, teacher or minister) could have been incorrect?"

> Why, if a man from Mars landed on [a reporter's] front lawn . . . [he] wouldn't be able to report his findings . . . because hundreds of people would be committing suicide and killing their children. They would have to admit that everything they know was absolute was not so. Yet they all are stupid enough to think that God—made all the billions and billions of stars and planets just for the few people on this one to sit and gaze upon.

Future Lifetimes

As skepticism mounted about "Bridey," Bernstein did have some good news. The British edition of the book was a big hit, and Bridey was on course to enjoy further "lifetimes" in other countries. Better still, there was a major feature film to look forward to.[60]

But the film was not the success Bernstein had anticipated. It entered production just as controversy about the book boiled over. Noel Langley, who wrote the screenplay and directed the movie, later described the experience as a farce. From the beginning, Hollywood's powers were nervous. They insisted that the climax of the film be "carefully constructed to scare the public off the irresponsible use of hypnosis as a party trick." They also demanded a scene in which a "protestant minister and a Catholic priest" warned that reincarnation was "subversive paganism." As filming progressed, Langley began to call his studio the "Contamination Ward." His production manager had even been told that the film would "never be finished"—but that Langley and producer Pat Duggan already were. When he asked to use the special effects laboratory to do fade-out shots, he found it was already booked for work on *The Ten Commandments*.

After *Life's* final dismissive article, the Bridey filmmakers found themselves moved from the "Contamination Ward" to the "Leper Colony." There was an effort to stop production entirely. In the end, Duggan and Langley did manage to finish the filming, but there was neither time nor extra footage for editing. The result was half-baked. "When it was released," Langley wrote, reviewers shunned it, and it "obediently wilted on the vine." Bernstein himself said that the "mess brewed in Chicago" killed his book and "murdered the movie."[61]

With the eclipse of the movie, Bridey's visibility declined still further. By 1960 she was gone from the national spotlight, but outside it the conversation continued. Virginia Tighe herself, reclusive and publicity-shy during the 1950s, embarked on a new Bridey-embracing career in the next decade. By this time the relatives she had feared to offend were senile, dead, or irredeemably alienated, and she could speak more openly. She often gave lectures to schools and church groups. In 1961, she mooted the idea of a new book on Bridey, and though she never wrote it,[62] she had other opportunities to tell her side of the story over the next few years in articles, a new edition of Bernstein's book, and media events. In 1965 she was featured on *To Tell the Truth*, a game show in which a panel of celebrities tried to identify the true figure among a number of false "challengers."[63]

As Bridey herself was fading from public discussions, an aspirational genre of parapsychological writing was emerging that inherited its defining features from her. It could be identified by its *In Search of . . .* titles. These books revived the John and Jane Does of past centuries. Bernstein's book itself remained the best known of the genre it spawned. It has been much reprinted, most recently in 2002 for the Book-of-the-Month Club. The Bridey experiments remained salient within this ongoing conversation about the possibility and nature of past-life regression. Bernstein himself continued to be part of it, and as recently as the 1980s he intervened in discussions of the Bridey case.

Past life "remembering" became part of a more general bundle of interests, all of them organized around the practice of age regression. Reincarnation itself became a perennially fashionable topic when the "uneasy fifties," as they have been called, gave way to an open, questioning era. The sixties produced a market for a dozen trade books on Edgar Cayce alone and supported the success of Alan J. Lerner's 1966 musical *On a Clear Day You Can See Forever*." During these years, the practice of hypnotic age regression by no means faded away. Rather, it grew and diversified, often in nonprofessional settings. It offered laypeople the prospect of getting at the truth about their own past experiences in circumstances when, for a range of reasons, these records were otherwise impossible to examine. It also offered a sense of robustness to claims by amateurs that circulated widely but were given the cold shoulder by professional and expert communities.

In November 1962, for instance, a couple traveling across the country found that they both seemed to have forgotten a significant period of time over the previous couple of days. They sought out Benjamin Simon, a World War II generation psychologist with expertise in age regression. With his help, they returned to the missing period, only to discover that they had spent it aboard an alien spacecraft. Thus began the era of alien abduction. Central to such claims was the conviction that age regression could revive these repressed or artificially blocked memories of extraordinary events.

As in the case of alien abduction, the memories brought back by age regression were always thought to be inaccessible because of a violent rupture of some kind. The original experience was so traumatic that it had been repressed from consciousness; or in the case of past life regression, death had intervened; or the experience might even have been deliberately blocked from future memory by an action on the part of someone else. In each case, age regression provided not only the means and the medium by which the memory was brought back, but in some cases the goal itself. The issues that had featured so centrally in the Bridey controversy—issues of why and how far to trust unusual or extraordinary testaments of memory, and how to contest or defend them—therefore figured in a proliferating range of controversies. They led, perhaps inevitably, to the fundamental question of how to carry out an experiment on a human being's mind and behavior. Anxieties surrounding the power of suggestion, wishful thinking, and subtle influence became central to the question of how any researcher could carry out research on a human being when that person's thoughts and behavior were part of what was to be studied.

6

SECURING MEMORY
IN THE COLD WAR

Out of the heady world of midcentury recall practices—trance-evoked experiences of battle, CIA interrogation research, and movies in the brain—came one that had significant consequences in American society. Those consequences would endure after the practice itself had receded from the public spotlight. The practice was forensic hypnosis, a technique that trawled a witness's mind for hitherto inaccessible records of the past, with the aim of producing evidence for a legal case. In the 1950s through the 1970s, forensic hypnosis grew from a little-known oddity to a well-known resource with unbounded therapeutic and forensic potential. By the late 1970s many police, lawyers, psychotherapists, witnesses, and defendants had come to see it as a golden key to unlock an otherwise inaccessible past and make it clearly visible to the institutions of the law.

The broad ambitions of forensic hypnosis were by no means new. Almost from its earliest days in the mid-nineteenth century, adherents claimed that hypnosis had the power to reveal legally significant truths hidden in the mind. But these efforts remained scattered and speculative, shunned by the courts.

After World War II, psychotropic practices had a new legitimacy. Science itself had achieved immense prestige in the wake of the Manhattan Project, and scientific ventures of all kinds were benefitting. Psychiatrists were particularly successful during these years, not least because of a boom in new practitioners who had been drawn into the field during the war. This new generation had used hypnosis and psychotropic drugs and saw these techniques as a straightforward part of medical practice. This new acceptance was institutionalized in 1958, when the American Medical Association conferred formal legitimacy on hypnosis as a medical practice, and psychiatrists founded professional associations centering on it. Psychiatric practitioners of hypnosis looked forward to a new era when it would unlock the secrets of mind, memory, and influence. They jealously guarded its fragile legitimacy, nervous about the tainting consequences of inappropriate practitioners without medical credentials, or about inappropriate applications beyond mainstream therapeutic purposes.

Psychiatrists did sometimes find themselves using hypnosis in legal cases, when a witness's or defendant's mental state seemed to fall within their areas of expertise or when a patient became involved in a case. Roy Grinker, John Spiegel, and William Sargant were all involved in forensic work in the postwar years. A less famous, but perhaps more representative, example is John D. MacDonald, who was a newly minted MD at the beginning of World War II and looking for psychiatric training in London. He found himself working under Sargant's direction in the Sutton Emergency Hospital, and for much of the war he used Sodium Pentothal on amnesiac and aphasic soldiers. After the war, MacDonald moved to Colorado and supplemented his medical practice by moonlighting as a forensic consultant. In most cases he evaluated people claiming insanity; in a few, he used Sodium Pentothal and hypnosis to get additional information about a crime under investigation, although he always warned that truth drugs could not coerce answers from unwilling subjects.[1]

If mainstream psychiatrists sometimes found themselves talking to lawyers or the police and using hypnosis in the course of this work, only few made this central to their careers. There were some mavericks who did. In the 1950s and 1960s, a few intrepid—some would say opportunistic—practitioners sought out lawyers and police officers, trying to create a market for hypnosis beyond medical treatment. They hung out their shingles, inviting attorneys to bring defendants and witnesses into their consulting rooms to "refresh" their memories and offering other hypnosis-related services.

By 1960 a nascent movement had begun to establish forensic hypnosis alongside the other instruments of judicial investigation and resolution. It gathered momentum over the next two decades, particularly after 1968, when a legal ruling opened the door to courtroom testimony based on hypnotically refreshed memories. By the late 1970s, tens of thousands of American police would have been trained in the use of forensic hypnosis, and in the understanding of memory that went with it.

William J. Bryan's Hypnotic State

William Joseph Bryan was the most conspicuous researcher on hypnosis and mental influence of his generation, and probably the most controversial. Great-grandson of the orator William Jennings Bryan, he had begun his career in World War II as a military psychiatrist. After the war, he is thought to have been involved in early research by the OSS (the organization that preceded the CIA), and later MKULTRA, the CIA's mind control research program, where he developed techniques of what he called "hypnoconditioning." I have found no solid evidence of government-funded work after the war, but Bryan's published writings certainly mark out an interest that straddles the forensic and the military range of psychological research.

The anxieties of the cold war accentuated the promise and peril of hypnosis for Bryan and his associates, evoking some of the central themes and concerns of the 1950s. The perceived threat to free will and individualism posed by Communism called into question how citizens could be both masters of themselves and socially connected to one another, a theme that was central to how he wrote about hypnosis: hovering behind his writings of the late 1950s was talk that five-year-olds were being brainwashed by Communist kindergarten teachers during snack time, that mental "zombies" were walking the streets (having been hypnotized by a chance experience), and more mainstream warnings that subliminal cues in films and advertising graphics were propelling viewers to the sales counter. There was also much talk of the threat of psychological research in Communist countries—that Soviet psychologists had realized that individuals' free will and personal identity were not essential and ineradicable but could be remade by experts. American psychologists needed to respond by understanding how the self could be broken down, shored up, or reconstructed entirely. Could one construct a new personality by destroying the existing memory and substituting a "rewritten" past? Could prophylactics insulate the memory and personality from stresses? Could memories be compartmentalized, so that in different states of mind one became different individuals? The best-known term associated with these kinds of questions was "brainwashing."

Forensic hypnosis was in some respects the opposite of brainwashing, at least in Bryan's formulation. It assumed that memory records were usually stable and permanent. It offered to sharpen the conscious mind's focus on its own memories, thereby strengthening the individual. This is not to say that brainwashing and similar research necessarily conflicted with forensic hypnosis: some people claimed to be experts in both. Both communities saw the mind as an open field for exploration and cultivation, but forensic psychologists relied on the stability and reliability of the memory record, while brainwashing practitioners relied on its malleability. This apparent conflict was not evidence of a clash between two clearly articulated beliefs about mind but rather reflected positions on a continuum of tacit psychological frameworks, along which practitioners moved depending on the kind of work they were doing. They emphasized memory's stability when carrying out forensic work and its malleability when speculating about or reflecting on brainwashing. Indeed, they could appeal to the apparent possibility of brainwashing to bolster the case for securing memories.

Bryan became interested in legal hypnosis in the late 1950s, when it achieved medical legitimacy and when, the following year, an obscure legal decision seemed to open the courtroom doors—just a crack—to hypnotists. In this case a defendant had hired a hypnotist to help refresh his memory of the night of a crime, and the tussle between his lawyer and the state over whether this was permissible was resolved in favor of hypnosis.[2] No decision was made about whether hypnotically refreshed memory could be admitted into evidence, but Bryan took the case as an invitation. He chatted up Los Angeles lawyers and police officers, attracting considerable interest. He struck it lucky with one in particular: Melvin Belli, a legendary plaintiff's lawyer now remembered as the "king of torts." Belli was then working to develop and promote the category of demonstrative evidence—physical and other evidence, such as videotapes, X-rays, models, and diagrams that are used to illustrate testimony but have no independent probative value. Belli became a champion of hypnosis, listing it in several books for tort litigators.[3]

F. Lee Bailey, another prominent defense attorney, described meeting Bryan in 1961 in San Francisco at one of Belli's popular series of one-day seminars, this one featuring Bryan as the main event.[4] Bailey was among a group of volunteers Bryan used to demonstrate his technique. He instructed them to hold out their right arms. He then described the arms as feeling stiff and numb. Bailey and the others thought they could feel their arms being rubbed as they waited for the next part of the show. Suddenly, at a command from Bryan, Bailey became aware that a hypodermic needle had been stuck through his hand. Bryan explained that this was an example of hypnotic anesthesia. But there was more: there was no blood where the needle had entered the skin. Bryan then told his subjects that they would soon see a slight drop of blood appear "in response to my suggestion." The blood appeared as he spoke.

Through these seminars and many other appearances at law society gatherings, Bryan amazed legal professionals with the power of hypnosis, and he soon built up a thriving consulting practice. He also launched a hypnosis consulting and teaching organization in Los Angeles, the American Institute of Hypnosis. The institute published a journal, issued textbooks, and ran training and research conferences.[5] Bryan himself also traveled internationally, delivering lectures, consulting, and staging teaching sessions at glamorous locations in America and Europe.

Belli and others also promoted Bryan's publications on hypnosis and forensic psychology. In 1962 Belli wrote an introduction to Bryan's textbook, claiming that hypnosis could gain access to "the learning, memory, and sensory mechanisms of the brain" of an amnesiac witness, drawing forth "facts" that had "slipped" below consciousness. And it did all this without disrupting the functions of the brain—a crucial reassurance for litigators worried about tainting witnesses' memory records. Al Matthews, a founder of the Los Angeles County Criminal Bar Association, was persuaded to hire Bryan as a consultant and recommended Bryan's course to other litigators. Matthews concurred with Belli that "hypnosis attains belief without doubt."

The key question for skeptics became whether this information was really true to an individual's memory. This was also the general issue for demonstrative evidence of Belli's stamp: whether it "fairly and accurately" depicted the thing it represented. Litigators worked hard to confirm that particular films or tape recordings could be trusted for the evidential purposes they had in view. Forensic psychologists, in turn, referred to these technologies not only because psychotherapists had done so, but also in the hope that the analogy would incline the courts to apply to hypnosis the legitimacy they were just then granting to tapes and film.

Manchurian Candidates

Alongside his legal and proselytizing efforts, Bryan also worked as a consultant to a number of Hollywood films and television projects. All of them involved specific and highly potent practices that could apparently alter

Advertisement for a film of Edgar Allan Poe's *Tales of Terror* (1962).

memories or personalities. In *Tales of Terror*, for instance, a Vincent Price film set in the nineteenth century, Bryan taught the actors eye contact techniques that he claimed were true to the practices of Victorian mesmerists such as John Elliotson and James Braid.

Bryan's projects also included, on his own account, *The Manchurian Can-*

didate (1962), a film that incorporated many of the psychological themes of his own career.[6] I have not been able to confirm Bryan's direct involvement, but the producer/director, John Frankenheimer, certainly drew on Bryan's favorite psychological works and themes.[7] The film itself hit on every memory fashion of the period, among them false memory, abreaction, dream memories, and multiple consciousness.

The Manchurian Candidate begins with a flashback of members of a U.S. Army patrol during the Korean War being captured and taken to Manchuria. In the present, a member of the patrol, Raymond Shaw, receives the Congressional Medal of Honor. Meanwhile another member of the patrol, Bennett Marco, has a nightmare that contradicts Shaw's heroic reputation: the patrol is sitting in a hotel lobby, listening to an old lady give a boring speech about flowers. As the camera pans across the room, the audience turns into Soviet and Chinese officials. The old lady turns into Dr. Yen Lo, director of the Pavlov Institute, who proudly describes conditioning techniques to control mind, memory, and behavior. Yen Lo instructs Shaw to kill a fellow soldier, and he calmly obeys. This dream sequence turns out to be a scrambled version of an extraordinary truth: the patrol was programmed with false memories, made to believe Shaw was a hero who saved their lives. Shaw himself was brainwashed more elaborately. A certain sequence of words places him in a trance in which he obeys all suggestions. He retains this information, yet cannot consciously recall it. Marco eventually uncovers a Communist plot to take over the American government using Shaw as an assassin.

The psychology at work in *The Manchurian Candidate* came from the Korean War brainwashing scare of 1950–53 and from mind control research later in the decade. In 1949 came the news that several American POWs held in North Korea during the war seemed to have been subjected to powerful mind control techniques. The power of this apparent "brainwashing" (as it was christened in a 1949 news article) was the subject of anxious debate throughout the 1950s.[8] Popular magazine articles announced that Communists were perfecting a combination of Pavlovian conditioning methods to achieve the "voluntary submission of people to an unthinking discipline and robot-like enslavement."[9] The government responded by funding research to develop home-grown techniques, including years of research on how to manipulate memory—enhancing it, extracting it, and "implanting suggestions and other forms of mental control."[10] D. Ewan Cameron, a Canadian researcher who performed some of the most extreme experiments, himself saw scientists as social engineers who could help bring about "a new world order."[11] Cameron's technique of "psychic driving" was aimed at breaking down personality and then reconstructing the individual's identity in a new form. Key to this was the erasure of autobiography and other forms of memory.

In the midst of this ongoing speculation about the real power of mind control techniques, some critics complained that Frankenheimer provided no account of how brainwashing worked.[12] In the novel, Richard Condon had specified drug treatments, hypnosis, and sensory deprivation; Frankenheimer did not want his film to get bogged down in details, so he merely had Yen Lo refer to the authors of the books Condon drew on more overtly, such as Andrew Salter's *Conditioned Reflex Therapy* (1949) and Fredric Wertham's *Seduction of the Innocent* (1954).[13] This lack of specificity might actually have boosted the film's credibility by failing to provide a target for skeptical evaluation. It also allowed future audiences to fill in the blanks with updated fears and fantasies, in a way that recalls Mary Shelley's silence about how Frankenstein's monster was created.

Even if he offered no causal explanation, however, Frankenheimer did have a view about what brainwashing was. He liked the populist theories, promoted in the press by Edward Hunter, which portrayed it as an irresistible, implacable force.[14] There were alternative accounts, perhaps less film-ready, that he could have chosen. For Eugene Kinkead, brainwashing was part of an argument about American degeneracy: it worked only because Americans were weak.[15] Other writers saw it as just one point on a spectrum of social and psychological influences. Social psychologists Martin Orne and Robert Rosenthal were discovering that scientists and experimental subjects could fool themselves and each other through subtle forms of communication and wishful thinking on a continuum with the "hidden persuaders" that sociologist Vance Packard claimed operated on every unsuspecting consumer. In this view brainwashing was not unique or even extraordinary. Frankenheimer preferred the theories of an irresistible drug-induced replacement of values and memories.

In fact, the many kinds of memory at work in the film presented a survey of contemporary recall practices, but with very different relations to truth and trustworthiness than one would conventionally expect. There is documentary memory, in the form of the apparently factual sequence that opens the film.[16] There is overt memory—that is, ordinary, on the face of things autobiographical memory—which here has a status very different from its usual role in films. We soon realize that this kind of memory is the one thing that definitely cannot be trusted. There is also not a repressed memory, but a conventional film flashback, when Raymond describes falling in love with Jocie when he was younger. It is oddly hokey compared with the darker tone of the rest of the film and feels strangely artificial, perhaps a sign of how different today's Raymond is from the young man of the flashback. There is dream memory, which we realize is somehow closer to the truth than conscious, "ordinary" memories. (The dream sequences are some of the film's most technically impressive features and some of the most famous in the film.) Finally, there is psychotherapeutic memory—autobiographical reflection on the analyst's couch.

These subtle ambiguities about fact and fiction pervaded not only the plot and structure of *The Manchurian Candidate*, but its meaning as a document. Contemporaries argued about whether it was a political satire (on McCarthyism) or a docu-warning-thriller; a dramatic story or a demonstration of real techniques. More recently, it has been the focus of speculation about whether the plot of the film could have played a role in inciting the presidential assassination carried out while it was still playing in American theaters. Lee Harvey Oswald often passed a cinema where *The Manchurian Candidate* was playing in the weeks before the Kennedy assassination.[17] The film even came to represent the reality of brainwashing as experienced by those who had suffered the technique. In later years the opening scene would be used in historical documentaries, presented as if it were taken from a documentary or research film instead of a work of fiction.[18]

In this context of uncertainty and possibilities, Bryan himself acted as something of a bridge between fact and fiction in a world of Manchurian candidates. In 1962, his journal warned that Soviet scientists

> have discovered a new process of relatively rapid thought control which they are using to mould men's minds to do their bidding. An example of this was revealed not long ago in news bulletins which spoke of the case of British spy, George Blake, who had been a prisoner of war in Korea and was apparently powerized (a new powerful method of hypnotic thought control), sent home to Britain with the subconscious suggestion firmly planted in his mind that he would deliver the information which he gained about his work over to Russian agents, which he dutifully did for a number of years.[19]

Bryan defined "powerization" as a kind of superhypnosis and likened Blake to Raymond Shaw. Although hypnotized persons would not do anything against their moral code, powerization could change the code. One could insert a "post-hypnotic suggestion" deep in the "subconscious mind," where it would remain undetected until called forth by some signal. Someone calling herself Edith Kermit Roosevelt expressed similar fears in a report that claimed the well-known psychologist Leonard Carmichael had warned of the development of a "scientific procedure that would allow the direct manipulation from the outside of the wellsprings of human behavior."[20] Whatever their nature, psychological techniques being honed in Communist countries were exceptionally dangerous, Bryan maintained—"far more dangerous than the atomic bomb." After all, "when you control the mind of the man who pushes the button, then you control all the bombs, all the weapons." He cited as evidence a brainwashing primer that he claimed had been written by Lenin himself for use in America, outlining the use of psychiatry for social control.[21] The point was that the United States needed a "crash program for thought control," on a scale that would make the Manhattan Project look like "peanuts."

Murder by Film

Bryan's own career brought together the fictions and facts of the world of *The Manchurian Candidate* in a series of often creepy, sometimes campy, and occasionally unbelievable projects. They began in 1960 when he was hired as a consultant in a high-publicity serial murder case. Henry Adolph Busch, a Hollywood optical technician, had confessed to three murders after being picked up on a charge of purse snatching. He claimed the film *Psycho* had made him do it: after seeing *Psycho* on September 4, 1960, he had been driven to commit his first murder (of sixty-year-old Shirley Payne), and after reminders of the film's notorious shower scene he had killed two other women. The prominent defense lawyer Al Matthews took on Busch's case, partly as a chance to probe the legal possibilities of Bryan's forensic techniques. Matthews wanted to try out a promising but controversial line of defense.[22]

Busch was not eligible for a traditional insanity defense, in which the defendant could not distinguish between right and wrong. Busch knew his actions were wrong but claimed he had not been able to stop himself. Matthews wished to apply instead the doctrine of "irresistible impulse," an increasingly familiar but highly controversial convention.[23] He maintained that Busch was "innocent of criminal intent because he was still under the influence of the movie horror story."[24] He hired Bryan to regress Busch back to the night of the first murder. Under hypnotic age regression Busch relived the period surrounding each of his crimes. *Psycho* was in theaters then, and he had seen it several times and become preoccupied with the shower scene. His first murder was carried out in what Bryan called an "accidental trance state" after the first viewing; later viewings of the shower scene "reminded him of E.M. [his first victim]." Like the tragic main character of *The Manchurian Candidate*, Bryan argued, Busch had carried out his attacks in a trance. Busch, like Raymond Shaw, had had a trigger—in this case, *Psycho* itself or events that recalled the famous shower scene.

The jury did not buy Bryan's argument. Busch was convicted and executed.[25] But even though Busch's story did not persuade, his murders nevertheless became linked to *Psycho*'s reputation as a film of extraordinary emotional power. As late as 1969, Hitchcock himself mentioned the case in an interview with the *New York Times*. He had thought Busch claimed to have been motivated by the film only once, and he said that his response had been, "Well, I wanted to ask him what movie he had seen before he killed the *second* woman."[26]

Soon after the verdict on Busch came another case in which Bryan would play a more persuasive role. F. Lee Bailey took on the brief of Albert DeSalvo, prime suspect in the Boston Strangler serial killer case. The case reminded Bailey of when he met Bryan and heard Bryan's description of the Busch case: a series of obsessive killings with a sexual theme. The question was not how to amass enough evidence to make a case for his guilt, but to decide whether his confession (freely given) could be trusted. To that end Bailey commissioned three people to help: Robert Ross Mezer, a psychiatrist, Melvin Belli, and Bryan himself.

Bryan took forceful charge of the situation. He was a huge man, both physically (he weighed over three hundred pounds) and, as we have seen, socially. In the opinion of contemporary commentators and historians, his mere physical presence helped him "overwhelm" DeSalvo.[27] His theory was that DeSalvo was motivated by a displaced wish to kill his daughter, Judy, who had a disability that required painful massage treatments. It was DeSalvo's job to carry them out. Bryan speculated that he had had to "hurt" women in order to "help" them, in keeping with his agonizing treatment of his child.[28] Bryan's questions drove suggestively toward this conclusion. "Each time you strangled, it was because you were killing Judy, wasn't it?" he demanded. "You were killing Judy."[29] In the presence of several witnesses, DeSalvo—closely questioned and guided by Bryan—described entering a victim's apartment, repeated what she had said to him (in an imitation of her voice), and then described the murder. This was not an unbroken narrative but an account pieced together by several questions. Bryan brought him out of the trance only after giving him a posthypnotic suggestion that he would have an informative dream that night, presumably with the hope of bringing some of DeSalvo's (apparent) memories of the murder into his conscious mind. The next day, DeSalvo did indeed describe a dream in which he killed the victim in question.

> DR. BRYAN: "Judy is going to be all right. *But you have to hurt her before you can help her*. [Quoting DeSalvo's description of a dream:] 'Don't scream. I won't hurt you' and she said, 'Okay.'"
> DESALVO: "No."
> DR. BRYAN: "What did she say?"
> DESALVO: "She can't talk. She's only a baby."
> DR. BRYAN: "She can't talk, she's only a baby. In other words, if these women were really going to be identified with Judy, the way they should be, they couldn't talk. Is that it?"

> He paused. "How could you keep them from talking? Come on, how could you stop them from talking." A pause. "Strangle them! That's true, isn't it? Isn't it? That's why the strangling."[30]

The detectives who heard the tape of this entranced interrogation were impressed. Bryan seemed to have unearthed new information that DeSalvo could not have learned from what was generally known about the crime. Bryan's work therefore helped strengthen the view that DeSalvo was indeed the Strangler. Bailey negotiated a plea bargain to avoid the death penalty, and DeSalvo received a life sentence. He was murdered in prison in 1973. But there would continue to be waves of controversy over the case as new claims of DeSalvo's innocence were made every few years, followed by rebuttals reasserting his guilt.

The Boston Strangler case was unusual in its prominence and in its bevy of heavyweight experts, but it was not unique in its use of forensic hypnosis. During the 1960s these techniques were used with increasing frequency in other cases, large and small. For instance, in 1962 Arthur Nebb of Columbus, Ohio, carried a loaded gun past several witnesses, walked into his estranged wife's house, and emptied the gun into her and her lover. The lover died, the wife was severely injured, and Nebb was charged with first-degree murder. At trial he claimed amnesia, and his attorney proposed using forensic hypnosis—on the stand itself. The judge was intrigued: he sent the jury out and allowed the experiment to proceed. The entranced Nebb—who, it is worth noting, had already been hypnotized on several occasions before the trial—now produced a complete narrative. He had not intended to use his gun, he recounted, but when he saw his wife with her lover he had gone into a trance. The gun had gone off unintentionally, and he continued to pull the trigger involuntarily until it was empty.

The performance of testimony here was complex. Nebb's attorney suggested that hypnosis not only refreshed Nebb's memory but validated the account that it made manifest. Hypnosis was a state of automatism in which one could not censor oneself. The validity of the hypnotic trance in the courtroom therefore gave legitimacy to the power of that other trance—the one that Nebb, while his will was vitiated in his current trance state, described as having vitiated his will during the shooting itself. It worked. In fact, from the historian's perspective it worked all too well: Nebb's performance was so convincing that we are left with no official record of that performance. After seeing the hypnosis session, the prosecutor decided to settle for second-degree manslaughter and a prison term of one to twenty years, with the possibility of parole in eight months. The deal meant that no detailed official record of the hearings would be retained. All that are left to us are newspaper articles.[31]

Similar incidents multiplied in America's courtrooms. The 1960s saw a series of cases that repeatedly tested the status of statements made using consciousness-altering tools like Sodium Pentothal and hypnosis, and of the experts who administered them. The legal door was opening to the use of forensic trance techniques.[32]

The Culture of Forensic Hypnosis

By the late 1960s it was easy to learn forensic hypnosis. There were introductory textbooks and recorded materials, weekend courses for beginners in major American cities (particularly Los Angeles, Las Vegas, and New York), and refresher or advanced courses at holiday destinations.[33] Police departments hired consultants to teach their detectives the craft. More advanced practitioners could hone their skills at specialized events. Bryan's own institute was one of a few that offered formal training. Enthusiasts could also sign up for one of many weekend training conferences in glamorous locations like Hawaii, Paris, and coastal Mexico. After teaching sessions, trainees schmoozed at glitzy cocktail parties, formal dinners, and ballroom dances.

The world of forensic hypnosis was full of imaginative devices. Of course there were spirals and pendula—the usual stuff of hypnosis. But one could also buy a "Thermamassage" chair that heated up and vibrated, supposedly encouraging the relaxed concentration conducive to the trance. Bryan himself was particularly enamored of a machine he used to manipulate memory electronically. The Schneider Brain Wave Synchronizer produced blinking lights at adjustable speeds, like a metronome with the sound turned off; in Bryan's words, it acted as "a photo-electric instrument that tunes to the natural frequency of the brain." He claimed that the synchronizer deepened the hypnotic "level" and increased the number of people who could attain the hypnotic trance.[34]

This device had been invented by a radio and psychology enthusiast, Sidney A. Schneider. During the war, Schneider had to check that American ships were properly equipped with radar. He observed that radar operators sometimes seemed to enter a trance while they were working, and he speculated that the pulses of light from the radar apparatus were somehow "synchronizing" with their brain waves. He introduced his Brain Wave Synchronizer in 1958 to reproduce this effect commercially, to help more people plumb their subconscious memories in a hypnotic trance.

Schneider Brain Wave Synchronizer, 1960s. Courtesy of Larry Minikes.

The device promoted one view of the relation between brain waves and the psyche: the idea that particular patterns of electrical activity were tied to particular states of attention and consciousness. Only if this were so could one alter states of consciousness by forcing the brain's electrical activity to change in synch with the light pulses of a machine calibrated to a particular state of mind.

Another claim about brain waves and memory, however, starkly conflicted with this. In recommending hypnosis to lawyers as a way of refreshing memory, Melvin Belli assured them that it would leave the "patterns of the brain waves unchanged and undisturbed." His reassurance appealed implicitly to the idea, widely credited in these years, that a particular pattern of "brain waves" was a defining feature of personal identity. The 1953 Hollywood film *Donovan's Brain* provided a sensational version of this notion: after a man's brain is removed following death, an EEG shows that it continues to emit brain waves. The brain subsequently takes control of the researcher through telepathy and attempts various crimes.

It is clear from these two understandings of the relation between brain waves, personal identity, and memory that forensic practitioners did not subscribe to a common or consistent account of what they were doing. But there were commonalities. Many did liken memories to "tape recordings." They usually subscribed to a physiological, reductionist account of cognition, or else to a mechanics of attention. "There can be no twisted thought without a twisted molecule," one rather literal-minded article declared. Another paper provided a diagram of the mechanisms of hypnosis. Bryan himself made gestures toward a kind of Enlightenment-mechanics system of hypnotic forces. In one case, he and colleagues used a scattershot pattern of little dots to represent "thought units" that were apparently disorganized during ordinary states of consciousness. On this account, hypnosis acted as a lens to organize them.

On the other side of America there was another hypnosis institute, controversial not for its leader's flamboyance and controversial claims about mind (or perhaps not only for these things), but because he had no medical degree. Harry Arons of New Jersey was a "lay" practitioner and a champion of "lay hypnosis," a practice intensely disliked by psychiatrists, who were trying to cleanse hypnosis of its stage associations and make it into a respectable branch of medicine; a distaste for lay practitioners was probably intensified by the debate, then current among the same community, about the practice of psychotherapy by people without medical degrees. Arons was shunned by most psychiatrists, but litigators and police had no concerns about his

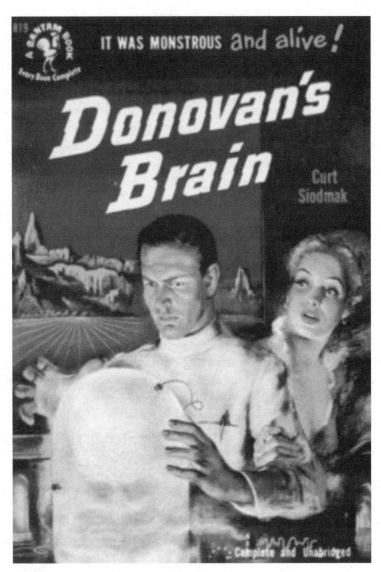

Advertisement for the film *Donovan's Brain* (1953).

credentials. He carried out a thriving consultancy and teaching program in New Jersey in the 1960s. His hypnosis textbook cited work in the neurosciences on the permanent storage of memory. He quoted neuroscientist James Birren, who argued that permanent memory records were created by "minute structural changes in brain cells," and he cited Penfield's claim that memory records were laid out in the brain like a strip of motion picture film.[35]

The conflicting, even chaotic mix of claims did not trouble practitioners themselves, who rarely fought about them. It seems that the various explanations functioned as a smorgasbord from which practitioners chose whatever fit their tastes and practice. The variety of explanations makes sense in this context, since it would be counterproductive for these very pragmatic individuals to spend time arguing about different theories when they could just choose one that seemed to work.

Bryan and his associates saw hypnosis not as a discrete technique but as a set of ways of understanding and managing an individual's relationship with the environment. They saw the social world as already suffused with subtle mental influences and altered states of mind. Judging from letters to the editor, most of the readers of the *Journal of the American Institute of Hypnosis* were not military mind researchers involved in projects like MKULTRA, but the forum allowed them to feel as if they were and to bond with each other in a mutually reinforcing fantasy. They discussed the many news stories, films, and scientific researches that portrayed the human mind as adrift among many currents of incoming influence. There were British children who suffered seizures while watching television—a response attributed to the hypnotic effect of the television medium;[36] there were American children whose teachers needed to be screened for Communist sympathies. And there was a Boris Karloff horror film of 1963, *Dementia 13*, whose audiences had to take a test, designed by Bryan, to determine their "emotional stability" before they could buy tickets. People who failed supposedly were "asked to leave the theater"—though there is no indication that anyone ever did fail.[37]

Bryan revised the familiar diagnostic category of "traumatic neurosis" to keep it up to date. His sober new term was "walking zombie syndrome"— "zombie" because those in this state were emotionally dead, even though they were physically healthy.[38] They could recover only by revisiting the past and re-experiencing it in a healthier way. Bryan represented this as an extreme example of a widespread, even universal, problem that called for a new understanding of mind and memory. We all "go along" in a semitrance, he argued, continually moving back and forth between "waking" and "hypnotized" states. The difference between a waking and an entranced state was so subtle that it was impossible for a layman to tell them apart.[39] No one, it seems, embraced Bryan's terminology. But the underlying idea was widely held: that much of mental life and interpersonal influence took place without

Advertisement for the film *Dementia 13* (1963).

our conscious awareness, and therefore without our self-scrutiny or ability to intervene.

Meanwhile, police continued to invite hypnosis practitioners into their investigations occasionally in attempts to solve many kinds of problems.[40] There were some spectacular cases of confessions brought about by a recovered memory, but these were few and far between. In most cases hypnosis was used to retrieve the memories of the victim. In New Jersey, Harry Arons was responsible for such ventures. Arons promoted a reassuringly simple understanding of memory. "Scientific research tells us," he wrote, "that we retain most of our experiences in our subconscious storehouse of memories." There was a fine grain to the memory record, so that even the most insignificant event was "stored somewhere in the back of our minds." Under special conditions, one could "bring them back to conscious awareness." How hard this was depended on the memory. Painful memories were often relegated to a "deeper niche in the subconscious" and were harder to recover. But one did not have to bring back everything. Even scraps, like "a partially remembered license plate number," could make all the difference. If so, there was a crying need for forensic specialists who could do this work. Doctors were not appropriate for it, because their professional prejudices would lead them to "scare police officers out of making adequate use of these techniques." Instead, hypnosis should be taught at police academies themselves, built into the curriculum.[41]

As such recommendations indicate, memory recovery techniques were primarily understood as a collaborative effort between crime victims, or onlookers, and police. But there were exceptions, most often when witnesses were not aware that they themselves had committed crimes. For instance, a New Jersey psychiatrist hypnotized the amnesiac mother of a dead boy who had been stabbed many times and had his head crushed. The aim was to help her remember what happened. Under hypnosis, it emerged that the mother herself had accidentally dropped a heavy rock on her child's head, and when she realized he was dead she had mutilated his body to conceal her guilt. As in this case, when trances did generate confessions, most cases involved some kind of mitigating contextual information—this mother had not intended to kill her child, and her distress was the reason for the amnesia. In a few cases confessed criminals actually requested memory refreshing help so they could cooperate more fully with the police and, presumably, negotiate more lenient sentences.[42]

At least one lawyer extolled the importance of these techniques not merely as memory refreshers, but because they could make courtroom testimony more convincing. Hypnotic review of memories made witnesses more confident about them. And hypnotists could make posthypnotic suggestions to wobbly witnesses to shore up their stories. One such witness was an inconsistent wreck before hypnosis, he noted, but afterward she shone at the preliminary hearing. She remained "relaxed and calm, remembered everything and answered all the questions coolly and correctly. She was not rattled under cross-examination and recross and was commended by the judge for doing an excellent job of 'taking hold' of herself and testifying with composure."[43]

In this description, the witness was the one in control. Hypnosis gave her a degree of self-command that helped her survive the cross-examination. Here, then, was a different understanding of the relation between trance, testimony, and memory than what one usually heard in popular accounts. It clearly drew on the idea that hypnosis could not make people do something foreign to their nature or character.[44] But it also involved a very different notion of the courtroom setting itself—one in which the court hindered or endangered truthful testimony rather than securing it.

When hypnosis was used on a defendant, the goal was usually to document intent, on the assumption that hypnosis could revive one's original state of mind during a crime. This was said to be an otherwise unmapped aspect of criminal activity. Bryan himself wrote in 1961 that in most cases, no attempt was made to explore the criminal's "true intent." He gave the example of a hypothetical suspect, "Charles Downs." When he is stopped for speeding and asked, "Where's the fire?" "Mr. Downs calmly produces a pistol and blows the police officer's brains out." Downs had committed no crime, had no obvious motive, and seemed "sane" in a preliminary exam: "Under hypnosis, however, we are able to uncover a reasonable explanation for Mr. Downs' behavior. In his childhood, we find that a burglar set fire to his home, shot and killed his father, and was going to kill his mother, when Mr. Downs, a small boy of six, shot the burglar in the blue suit." Mr. Downs came to hate the color blue so deeply that he was "on the verge of insanity," and kept a gun for self-defense. When the policeman "unknowingly integrated the fire with his blue suit," Downs became temporarily insane. He intended to attack not the *policeman*, but the suit. Only an expert using an appropriate technique could discern Downs's true "intent." Bryan's point was potentially quite consequential. The interest of law schools, district attorneys, judges, and defense lawyers in such techniques told him there would soon be changes in criminal charges, not just the ways defendants could rebut those charges.[45]

The key claim of the forensic hypnotists—that they could scientifically "refresh" compromised memories to their original state—remained contentious, however. It assumed both permanence and stability in memory records and was far more sanguine about the representational character of memory itself than courtroom lawyers experienced in dealing with memory's foibles tended to be. Perhaps for this reason, courts resisted attempts to admit the results of forensic hypnosis well into the 1960s. Under the radar of official record keeping, however, police and litigators were by this time routinely using it in their everyday investigations. Police departments began to hold training sessions about 1960, even though they knew that hypnotically facilitated statements could not (yet) be entered as evidence in a criminal proceeding.[46] The forensic hypnosis entrepreneurs profited. In 1964 Bryan's institute made a "Big Move" to larger premises.[47] On the other side of the country, Harry Arons's consulting service also expanded.[48] Arons reflected in 1967 that, early in his practice, the chair of the New Jersey Bar Association ethics committee complained that Arons was "placing in the hands of policemen a tool by means of which they might be enabled to force confessions from innocent people in their zeal to obtain conviction." And on another occasion, a course that had been organized with the approval of local citizens and police was canceled by the county prosecutor "because of fears that defense attorneys would accuse the criminal justice workers of "employing questionable methods." But since then his work had got traction. He became the chief investigator for a U.S. Air Force project involving forensic hypnosis, for example. Little is known about this project, but it was probably focused on extracting information from traumatized pilots of downed aircraft. His training courses and professional association flourished. By 1967 hundreds of criminal justice professionals had taken his courses, along with investigators from the Internal Revenue Service, insurance companies, and private detective agencies.[49]

As the programs in California and New Jersey flourished, others sprouted up around the country.[50] Invited lectures at universities and police departments became more common. One visiting lecturer at Case Western Reserve University, speaking to a packed room at the Law-Medicine Center in 1960, illustrated the rhetorical power they could draw on. Dr. Dezso Levendula described the difficulty of helping victims remember details of traumatic events and then, in the middle of the lecture, pulled the old trick of staging a fake crime in the lecture theater.[51] The students—all patrolmen—were asked to give accounts of what had occurred, which inevitably were full of holes and

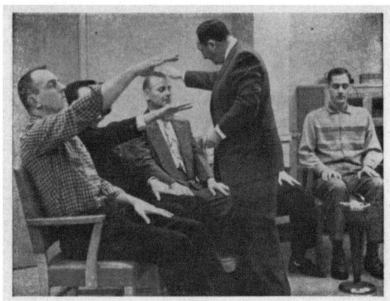

Dr. Dezso Levendula hypnotizes five policemen who witnessed a staged "murder" at the Cleveland, Ohio, Law-Medicine Center. The experiment demonstrated hypnosis can help eye-witnesses recall details of a crime

Demonstration of forensic hypnosis at Case Western Reserve University, 1960.

inconsistencies. Several students then submitted themselves to a hypnosis experiment.

A consultant for the Department of Defense then continued the lecture, explaining that one of the great virtues of hypnosis was that it created an intense, intimate bond with the subject. It was an extreme version of what police routinely sought to achieve in any effort to solicit testimony, and therefore a natural extension of investigative practices.

In 1968, forensic hypnosis finally achieved a degree of formal legal recognition. The occasion was a Maryland rape case. The victim, who had had been left for dead, could remember nothing. Police called in a forensic hypnotist, and she produced a complete narrative of the rape. The judge ruled that forensic hypnosis was sufficiently established as a scientific technique to justify admitting her testimony—though the jury could then decide to give it less weight than a statement made from a memory that had never been impaired. The defendant was convicted almost entirely based on the refreshed memory testimony.[52]

This decision threw the courtroom doors open to forensic hypnosis, making it one of the great technological hopes of American police departments and prosecutors.[53] Although only a small fraction of cases involved courtroom testimony based on refreshed memories, the techniques were nevertheless becoming part of police and jurisprudential culture. There was talk of a new age of investigative and testimonial certainty. It was as though the ambitions of Hugo Münsterberg, whose hopes for forensic psychology had been dashed more than half a century earlier, were finally being realized.

Yet it was not quite the final vindication of Münsterberg's original objectives. He had called for a unified body of experts with common intellectual assumptions. What in fact emerged in the sixties were discrete communities that often disagreed in fundamental ways. Psychologists and lawyers joined forces on a broader front during these years. They formed professional societies, journals, and programs at funding bodies dedicated to forging a nexus between legal and psychological research. Research psychologists, clinicians, and later neuroscientists were all drawn into their efforts. They ushered in a cooperative (albeit still vexed) relationship between the legal and the psychological fields that has continued to develop and intensify. The forensic hypnotists certainly aspired, and it seemed successfully, to be part of this emerging alliance. But fundamental inconsistencies of explanation, interpretation, and practice remained. And because of them the authority of forensic hypnosis remained fatally fragile.

The Rise of "Confabulation" and the Downfall of Forensic Hypnosis

On July 15, 1976, two men hijacked a Chowchilla, California, school bus carrying twenty-six children, took it into a deserted quarry, and forced the passengers underground into a buried moving van. After twelve hours of terrified solitude in the dark, they found a way out and were rescued.[54] As the search for their kidnappers got under way, it proved difficult to obtain any useful identifying information from the children or the bus driver—the kidnappers had worn hoods, and everyone's memory of specifics was patchy at best. The police called in well-known Los Angeles psychiatrist William S. Kroger, in hopes of using hypnosis to pull some useful leads from the memory of the driver, Ed Ray. Kroger sat with Ray in a Fresno motel and coached him in deep breathing until he was in a trance. Kroger then led him through what the papers called a "playback" of the kidnapping, whereupon Ray was able to recall all but one digit of the license plate of the kidnappers' van. The fleeing suspects were caught and convicted.

Although it occurred almost a decade after the Maryland rape case that opened the legal door to forensic hypnosis, the Chowchilla kidnapping has been called the catalyst that placed the technique in the public spotlight. By the time police enlisted Kroger, the horrific story was already making national headlines. So journalists were paying attention when Kroger seemed to reach into Ray's mind with surgical precision, pulling out specific information that was confirmed when police found the van and could see the license plate for themselves.

Kroger was one of ten experts working at a bespoke forensic hypnosis program run by the Los Angeles Police Department in its Behavioral Sciences Services. The head of the program was Martin Reiser, whom the LAPD had hired in 1968 as a full-time police psychologist (one of the first of his kind). Reiser became a prolific proselytizer for police psychology throughout the 1970s. Captain Richard Sandstrom, who had been charged with evaluating the unit's work, confirmed its impact, telling journalists that "hypnosis gives utterly fantastic results." The Los Angeles police who received Sandstrom's training were nicknamed the "Svengali Squad."[55]

But in the wake of Chowchilla, as even more police sought such training, public celebrations of forensic hypnosis lit a fire under a community of psychological experts with a very different perspective on memory.[56] Most cognitive and social psychologists remained far less sanguine about the effectiveness of fishing for pristine memory records, particularly using psychotropic techniques. They saw memories as constantly in flux. Social psychologists in particular had created a well-developed literature on the role of what one would popularly call "suggestion" in psychological experiment. They argued that the so-called memories vaunted by the hypnotists were a mixture of memory and suggested fantasy and that no one could tease them apart. Some of these psychologists now challenged the forensic hypnosis advocates head-on, in a series of sustained battles with the future of the field at stake.

Social psychologist Martin Orne was particularly active in sounding an alarm about the unreliability of statements secured by hypnosis. Orne had built his entire career on this kind of skepticism. During the years when Roy Grinker was evaluating William Heirens and Morey Bernstein was taking Virginia Tighe back to 1820s Cork, Orne was studying how experimenters could unwittingly "convey" to their subjects "an experimental hypothesis," and in so doing become "significant determinants" of their behavior. He analyzed how they imperceptibly encouraged subjects to respond—not to the stimulus that was being tested, but to subtle hints about the experimenter's hopes. The reaction was in a sense the counterpart to one that has later become more famous—Robert Rosenthal's "experimenter effect." Rosenthal's effect is popularly defined as "seeing what you expect to see." Rosenthal articulated it at the same time as Orne and partly in coordination with him, and it was publicized in tandem with Orne's work. Orne's counterpart could be defined as "subjects acting as they think you expect them to."

The idea that expectations could unwittingly produce a desired effect was by no means new. Famous episodes in which such an effect was invoked date back at least to the eighteenth century, with the Académie des Sciences' discrediting of animal magnetism.[57] But in the second half of the twentieth century psychologists became formally concerned about it and developed prescriptions for how to identify suggestive design flaws in human subject research. When Orne was an undergraduate in the late 1940s, psychologists routinely recounted how experimental subjects might try to please their researchers and how researchers tried to keep them ignorant of an experiment's key hypothesis. But was it enough to refrain from telling subjects what one was trying to prove? Orne thought not. He had a family background in psychiatry and an immediate interest in another practice that made him an expert in manipulating attention: stage magic. The stage magician's success centrally depends on using subtle cues to control the audience's attention and expectations, and Orne later credited his experience as a magician with sparking his interest in similar cues in experimental settings. Even as an undergraduate, writing his BA thesis in the late 1940s, he was already pursuing the idea that experiments involving the human mind centrally involved subtle communications between experimenter and subject. The practice he focused on then was hypnotic age regression—a controversial but promising psychological technique.[58] As we have seen, regression therapists claimed that their patients were objectively reliving the past. Orne argued instead that they were attempting to please the hypnotist, albeit without any conscious effort at fraud. He carried out a series of experiments to demonstrate this, first for his BA, and then in his doctoral work in psychology at Harvard.

Orne used hypnosis as an extreme case of what he claimed happens in any psychological experiment: the subject is motivated to please the experimenter and picks up on suggestions about what the experimenter hopes will be said or done. He delivered a general argument along these lines in a highly influential paper of 1962, in which he coined the term "demand characteristics" for these suggestions. Anyone wishing to study the human mind—not only in hypnosis but in any human subject research—would need to study the nature of this social encounter. Overall, an experimenter could be thought of as entering into an implicit contract with subjects. Subjects agreed to tolerate various inconveniences and discomforts, and the experimenter in turn guaranteed their well-being. Once subjects stepped into the experimental space, they became unusually accepting of any instructions: "Just about any request which could conceivably be asked of the subject," Orne wrote, "is legitimized by the quasi-magical phrase, 'This is an experiment.'"[59]

This was where "demand characteristics" came in. Subjects were motivated to respond to hints—or what they took to be hints—about what kinds of behaviors would make the experiment a "success," and they delivered. To properly evaluate experimental results, therefore, one should take care to evaluate the experiment from the participant's point of view. But this was difficult precisely because the subject's motivation extended to a "pact of ignorance" with the experimenter. Often the subject figured out the goal of the experiment but protected the researcher from realizing this. And while the subject became supersensitive to suggestions from the researcher, the researcher, in turn, became unusually *insensitive* to signs that the subject possessed this knowledge. Orne's explanation for this "compliance" began with hypnosis, which he saw as an extreme version of what was going on in any experimental setting. Hypnotized people were powerfully moved by their relationship with the hypnotist, and by cues that expressed his expectations. People who stepped into the role of experimental subjects were, he contended, influenced in similar but less forceful ways.[60]

One direction in which Orne pursued this line of reasoning was toward lie detection—a popular research topic at the height of the cold war. He published a stream of papers examining how factors such as the subject's motivation or awareness of the proceedings could affect the detection of deception in laboratory tests.[61] But his interest in these questions was unusually complex and ambivalent, because unconscious "compliance" was a central organizing principle for so much of his work. Deception was not an aberrant practice undertaken furtively by problematic individuals but a fundamental element in people's interactions. It was part of the social furniture of every psychological laboratory.

Orne soon went hunting for "demand characteristics," revisiting some of the best-known psychological experiments of the 1950s and early 1960s. Other researchers began to write papers on the methodological issues Orne raised, and then they more routinely discussed demand characteristics in projects ranging from psychotherapy to child development and sociometry.

Eventually the term became a staple in the "methods" sections of psychological papers, though still used more in attacks on others' work than in explanations of one's own.[62] Eventually Orne's 1962 paper became a standard item on psychology course syllabi. But still the concept was primarily cited when it could be used to frame a criticism of an experiment that had already been staged.[63] Most psychologists can now define the term when asked but rarely have occasion to use it.

Orne's dream for a psychological field devoted to the psychology of experiment never matured. No such subfield came into being. One reason was that the social structure of experiments themselves changed. The dyad of researcher and subject was eclipsed by the rise of large correlational studies. These were not designed to avoid the problems of demand characteristics, but their sheer numbers arguably made the kinds of worries Orne articulated—the subtle interactions between one experimenter and a small number of subjects—irrelevant. They instead marshaled psychological phenomena over a large demographic field. Researchers invested in a kind of neobehaviorism, partly in hopes of avoiding the need for subtle analysis of the relationship between researcher and subject.

Where the older kind of analysis remained salient, even central, was in an area of psychological dispute far afield from the psychological laboratory. Orne's arguments by no means stopped hypnotic age regression (attacks on parascience rarely have that power), and many forensic hypnosis experts thought of themselves as practicing regression. But this gave him a field in which to operate. Beginning in the mid-1970s, Orne developed a celebrity consulting career as a hunter of demand characteristics and a debunker of psychological fraud in individual cases of supposed multiple personalities or recovered memories.

In the hugely publicized serial killing case of the "Hillside Strangler," for example, Orne was commissioned to do a psychological evaluation of suspect Kevin Bianchi. There was no doubt that Bianchi was the murderer; but the way he was acting suggested he planned an insanity defense, using a claim of multiple personality. Orne used a quasi-hypnotic experiment as a kind of double-bluff to trick Bianchi into "inventing" a new personality in a way that discredited his claims to being a legitimate multiple personality victim, and revealed him instead to be fiendishly deceptive. Bianchi was convicted, and he died in prison in 2002. The case was soon being treated as a lesson in the trickiness of insanity diagnoses. The BBC's *Horizon* program broadcast a powerful documentary on the case in two parts, the first detailing the case for Bianchi's madness with little to hint that viewers would later have reason to doubt the diagnosis. The second part then dismantled the credibility of the first, leading to a resounding demolition of Bianchi's supposed condition. The program was a convincing invitation to entertain deep concerns about the diagnosis of multiple personality in general.[64]

About the same time, Orne also testified for the defense in the trial of Patty Hearst. He argued that Hearst's situation as a hostage had so compromised her ability to make independent judgments that she had been placed in a continuous "consensual" state in which she could not have refused to commit the crimes she was charged with. She was effectively a "private in an army of generals." Once again Orne was evaluating historical testimony, viewing them through his accustomed filter of the malleability of human decision making in situations where an individual was placed in a compromised, "consensual" state by particular relationships of authority.

About the time of Chowchilla, Orne threw himself into a sustained battle over forensic hypnosis itself. He carried on his campaign in a string of court cases from 1975 through the early 1980s. In 1975, Jeffrey Alan Ritchie, a man accused of killing a child of two and one-half, was confronted with an overwhelming amount of circumstantial evidence and requested hypnosis to help him remember details he could not consciously recall. Under hypnosis he relived the experience dramatically, apparently indicting his wife for the murder and clearing himself. Orne led the court through the videotape of Ritchie's entranced testimony to indicate moments where, he claimed, the defendant had been led by the hypnotist. The court ruled to exclude the hypnotic evidence as unreliable. The same issue arose about the same time in the retrial of a convicted murderer in *State of Ohio v. Papp*, where the court had permitted forensic hypnosis of the defendant. He seemed to exonerate himself, and the Ohio press proclaimed his innocence. Again videotapes were available, and again Orne led the court through them, indicating moments where the defendant was dissimulating. He also suggested that the lay hypnotist hired by the defense should administer certain tests his laboratory had developed to identify dissimulation. These tests were done, and they convinced the court that the defendant had never actually been hypnotized.

Such deliberate cases of deception were rare; far more insidious was the problem of suggestion in cases where all parties were acting in good faith. After many interventions in criminal cases, in 1977 Orne made a formal prescription about the use of forensic hypnosis in an amicus curiae brief in the case of *People v. Quaglino*. Here he reiterated that hypnosis could be useful in producing new information but also produced "confabulations." Confabulation was a psychological term dating back to the early years of psychoanalysis, referring to a process that knitted memories of past experiences together with fantasies or suggestions to form an imaginative construction that appeared to be a memory. In this context it meant a constructive, imaginative filling-in-the-blanks in which incoming suggestions from an interlocutor played a key role. Confabulations were innocuous as long as they were always checked against corroborating evidence (and ignored if none could be found). But only independent corroboration could distinguish between real memories and confabulations. Invoking the Chowchilla case, Orne warned that "there are many instances when subsequently verified accurate license numbers were recalled in hypnosis by individuals who previously could not remember them; by the same token, however, a good many license numbers which witnesses recalled turned out to belong to individuals where neither they nor their cars could possibly have been implicated."[65] The problem was that no one was insisting on the independent corroborations that would guard against the use of confabulations, even as forensic hypnosis swept through police departments across the country and was even being taken up by the FBI. Even the most expert psychologist—someone like himself—could not rule out the possibility of sending out cues that would unwittingly lead to the creation of a pseudomemory.

Orne proposed several "absolutely essential" safeguards.[66] Only a specially trained psychologist or psychiatrist should carry out the experiment, and "he should receive a written memorandum outlining whatever facts he is to know, carefully avoiding any other communication which might affect his opinion." All interaction between hypnotist and subject should be videotaped. Before the experiment, the subject should narrate what he or she could remember, for comparison afterward. No one but the psychiatrist and the subject should be present before and during the hypnotic session. And finally, tape recordings should be made of police and psychological interviews with the witness before the experiment, in order to document that no cues were sent out in these conversations that might lead to statements surfacing under hypnosis. These were the minimum safeguards—necessary but perhaps not sufficient—to use with witnesses who might later need to testify in court about their memories. Orne wrote up his guidelines more formally in a number of papers and in a brief commissioned by the FBI, which was seeking reliable ways of getting more information out of witnesses by using hypnosis.[67]

The definitive formulation of these rules came in a 1980 case, *State v. Hurd*. Herbert Spiegel, a veteran clinical psychologist and hypnosis expert, attempted to refresh the memory of an assault victim who could not remember the identity of the person who attacked her in her home one night. During the hypnosis session she identified a candidate; but afterward she was reluctant to confirm the identification. Spiegel pushed her to do so, remarking among other things that her failure to make the identification would leave her husband on the list of suspects. Orne was enlisted as an opposing expert witness, and the two men clashed on whether it was possible to completely prevent the creation of false memories. The court believed Orne and accepted his guidelines for all legal use of hypnosis on witnesses.[68]

A crushing blow came two years later, in 1982, in a California rape case. The case was a terrible example of when and how not to use hypnosis. It was a standard he said/she said case, with consent the central issue. This made the stakes very high in terms of the evaluation of the hypnotically refreshed memory of the victim, who initially could not remember the event at all. The complaining witness had been drinking heavily, which on its own could have supplied the reason for her memory problems (as one of the justices put it, there was a "good possibility that she has no clear memory to be refreshed").[69] She was also using tranquilizers and had had psychiatric treatment in the past. And she was up against a defendant, Donald Lee Shirley, who was in the military and had excellent character references. Not only was the witness's own memory compromised (by the alcohol), the investigation was problematic, even in the eyes of some of the most stalwart supporters of forensic hypnosis. Her hypnosis was conducted at the courthouse by a deputy district attorney—not by a psychiatrist in a controlled location. And it was done the day before the trial was to begin. Despite these problems, and in the face of warnings from expert witnesses for the defense about the problems of hypnosis in this case and in general, Shirley was convicted—based solely on testimony recounting hypnotically refreshed memories. Shirley's appeal became the test case for hypnosis in California—and because of the importance of the California Supreme Court, for many courts elsewhere in the country.[70]

The debate soon focused on the per se admissibility of any posthypnotic testimony, not just whether hypnosis had been carried out appropriately in this particular instance. By this time there were fifty-nine closed cases that could be reopened if hypnosis were banned; and many more open ones stood to have their verdicts affected.[71] Psychiatrist Bernard Diamond of Berkeley, who was very influential in California psychiatric law, had written an explosive article in 1980 taking Orne's arguments to an extreme. All hypnosis affects memory, Diamond argued, regardless of any safeguards one might try to put in place. He listed fourteen questions he thought litigators should to ask when cross-examining a hypnosis expert, all aimed at undermining the credibility of a forensic test. His point was that the only safe course for the courts to take was the total exclusion of testimony from any witness who had undergone hypnosis.[72] Diamond's decision was put into practice in the Shirley case.

The case concluded on March 11, 1982, when the court made a sweeping exclusion of all use of hypnosis on witnesses. It condemned the arguments of Reiser and others that memories were similar to tape recordings, and stated that memories were "productive, not reproductive." It rejected the safeguards Orne had offered a few years earlier, on the grounds that even these did not adequately guard against confabulation and false memories. The decision drew a firestorm of controversy. Forensic psychiatrists protested that only the most naive hypnotists claimed memories were like tape recordings—most understood that suggestion could change memories, and there was widespread support for Orne's guidelines for how to avoid this problem. Others argued that since psychotropic practices were very commonly used in psychotherapy, particularly with people trying to recover from traumatic events, this ban would preclude prosecution in many current cases and prevent future victims from seeking the therapy that could help them. Police officers also detested the decision, believing that it removed control of crucial interviews from the people who knew the most about how to handle witnesses and suspects—the police themselves—and handed it to doctors who knew comparatively little about gathering and handling evidence.[73]

7

FLASHBULB MEMORIES

In 1977, two Harvard psychologists announced a new concept in the study of memory. It was inspired by something they thought most readers had experienced. "Almost everyone can remember, with an almost perceptual clarity, where he was when he heard" about the assassination of President John F. Kennedy, wrote Roger Brown and James Kulik: "what he was doing at the time, who told him, what was the immediate aftermath, how he felt about it, and also one or more totally idiosyncratic and often trivial concomitants." Everyone knew a great deal about the assassination, of course, but this knowledge did not interest Brown and Kulik. Rather, they were intrigued by people's memories of what they themselves were doing when they first heard the news. This kind of information had no obvious utility, and yet the memories seemed unusually informative, quite different from the abstract, hazy quality of most long-term recall.

Brown and Kulik thought something special was going on here. When people heard shocking news, particularly distressing news, their brains apparently recorded it in a way that was different from ordinary memories. These special memories were especially faithful to the original event and resistant to the usual fading and mistakes of regular long-term memories. It was as if the mind had taken a photographic snapshot of its own experiences at the moment of the news, preserving everything, regardless of relevance. They called such instances "flashbulb memories," to underscore the photographic reference. Over the next two decades their idea prospered, and so did the reference. Flashbulb memories came to be widely acknowledged, whether as "Kodak moments," "Polaroids," or more mundanely, "snapshots."[1]

Although Brown and Kulik thought this was a basic (and therefore long-standing) part of how the mind works, the kind of event that inspired their work was relatively new. Theirs was the first generation to receive news by television, and therefore to learn simultaneously about major events through transmissions that very quickly became a common pool of remembered experiences. This collective media experience was not itself a defining element of "flashbulbs," but it called Brown and Kulik's attention to the memory phenomenon, and it also supplied a means for studying it. Broadcast news created events that were experienced by many diverse audiences as a single shared moment of revelation. Most people would have memories of such moments, which could therefore be investigated by surveying large pools of citizens.

Collective Grief

The events that intrigued Brown and Kulik were part of a pattern more social than psychological: the kinds of conversations people had with each other about the Kennedy assassination. A decade after his death, it had become common for people to ask each other, "Where were you when you heard that the president had been shot?" The question was so standard that the first three words sometimes sufficed. During the publicity surrounding the tenth anniversary of Kennedy's death, a journalist had asked this of several celebrities. It turned out that Julia Child was in the kitchen eating *soupe de poisson*. Tony Randall was in the bathtub. Billy Graham was on the golf course but felt a "presentiment of tragedy."[2]

Brown and Kulik had their own vivid memories of the assassination. Brown was "on the telephone with Miss Johnson, the Dean's secretary, about some departmental business. Suddenly, she broke in with the news." Kulik was in his sixth-grade music class and heard the news over the intercom: "everyone just looked at each other. Then the class started yelling, and the music teacher tried to calm everyone down." When he got back to his homeroom, his teacher was crying. "They told us to go home."[3]

The assassination was a collective shock on an unprecedented scale. As a much later commentary put it, "Before 9/11, there had been the Kennedy assassination. Before the Kennedy assassination, there had been nothing."[4] What was unprecedented was not the assassination, but the new media environment.[5] The shared experience of listening to simultaneous radio programming had already created a new kind of community; in the 1950s and 1960s, television culture was built on radio culture. The proliferation of recording technologies that could produce traces of experience replayed in real time redefined what an "experience"—and a memory of that experience—was understood to be. One commenter reflected that television had the "knack of fading the present into the past, scene by scene in direct juxtaposition," which encouraged a "continuity of feeling" about past and present.[6]

Most people answering survey questions about the news of Kennedy's shooting claimed to have heard within an hour. A common set of images became universally associated with the assassination and its aftermath. "The midnight-blue Lincoln Continental, Jackie Kennedy's pink suit and pillbox hat" were shown over and over on television at the time and became "seared" into the memory of viewers.[7] The mutual awareness of not only living through the same transforming event but learning about it at the same time and through the same images inspired the poignant "Where were you . . . ?" question that became part of the nation's collective history.

What were the salient qualities that would qualify an event to become a memory of this kind? Did everyone respond with the same strength to the same shocking events? Or did memories vary in strength according to how much one cared about the news? Brown and Kulik decided to answer this last question by studying reactions to events that mattered more to individuals of one socioeconomic or racial group than to those of another. If such events produced "flashbulbs" more often for one constituency than another, this would support the idea that the emotional significance of the news, not just its novelty, was important. And the largest demographic divide, Brown and Kulik believed, was between black and white. So they designed a survey with questions about several different events, particularly assassination attempts on public figures—for instance, the assassinations of Martin Luther King Jr. and Malcolm X.[8]

There were a few surprises among their results. In general, the survey confirmed their expectations. Black people did form "flashbulb" memories more often than whites after the deaths of black civil rights leaders. Whites formed them more often after Chappaquiddick. But Kennedy's death was different: black and white informants reported the same number of vivid memories. Brown and Kulik took this to be a sociological rather than a psychological finding: they had underestimated Kennedy's importance to black people.

TABLE 1. Chronological order of events used to search for flashbulb memories

Medgar Evers	Black	Shot to death	June 12, 1963
John F. Kennedy	White	Shot to death	November 22, 1963
Malcolm X	Black	Shot to death	February 21, 1965
Martin Luther King Jr.	Black	Shot to death	April 4, 1968
Robert F. Kennedy	White	Shot to death	June 5, 1968
Ted Kennedy	White	Drowning involvement	July 18, 1969
George Wallace	White	Shot, but not killed	May 15, 1972
Gerald Ford	White	Failed attempt at assassination	September 5, 1975
Gen. Francisco Franco	White	Died of natural causes	November 20, 1975

Photographic Memories

Brown and Kulik named their finding for its "primary, 'live'" quality. It was "like a photograph," not only in its visceral power but because it "indiscriminately preserves the scene in which each of us found himself when the flashbulb was fired." They referred to the lighting technology that made the photograph possible rather than to the photo itself, because they wanted to convey the idea of "indiscriminate illumination" as well as "brevity." Flashbulbs were an evolving technology at this time; the huge contraptions of the early twentieth century had given way to powerful little cubes (Flashcubes, then Magicubes) or strips (Flipflash). Each bulb could flash only once, so the illumination was uniquely linked to a single photograph.

Although Brown and Kulik referred to still photography, their idea was influenced by several decades of home moviemaking. Amateur films had long been marketed by camera and film distributors as an external form of memory, and as we have seen, midcentury advertisements for cameras and film presented home movies as a way to relive the past with perfect fidelity. Penfield's work was indebted to this kind of claim. But Brown and Kulik's debt to home movies was more complex and indirect, by way of a single amateur film, the most famous ever made: the "Zapruder film" of the Kennedy assassination.

Abraham Zapruder shot the silent eight-millimeter film from his office overlooking Dealey Plaza, producing the most complete recording of the assassination and arguably the most reproduced, viewed, and analyzed home movie ever made. The film was developed by Eastman Kodak in Dallas hours after the event. Zapruder gave copies to the Secret Service for its investigation and sold print rights to *Life* magazine. *Life*'s November 29 and December 6 issues printed many frames. More were printed on October 2, 1964, November 25, 1966, and November 24, 1967.

The film itself, and the history of its reproduction and distribution, became central to claims about what really happened when Kennedy was shot. The film's fidelity to the events of those famous "six seconds in Dallas," the fidelity of stills to the moving image, the fidelity of sketches made from stills, and the fidelity of copies of the film to the original have been debated ever since.[9] Most immediately relevant is that when people spoke of their memories of the assassination, they were often thinking of the stills and short sequences published over and over in *Life*, and eventually everywhere else. The film has even become part of the memories of individuals too young to have had any experience of the original news, because excerpts have figured in so many films and documentaries. The point here is not merely that individual and collective memory became dependent on a particular home movie, but that the status of amateur film—both in general and in the specific case of this one film—was that of an experiential record, more trustworthy in its relation to past events because it was not made by a professional. The imperfectness of these films—their poor composition, shakiness, and so on—was a hallmark of a lack of "art" (in the sense of artifice), and therefore a sign of veracity.

Flashbulb memories were also part of a mode of cultural remembering that was defined by the photographic "moment"—a moment of experience that was thought to express the essence of something much bigger than itself. When still photography was introduced in the mid-nineteenth century, it was immediately likened to memory, and vice versa. But photographs of the 1840s and 1850s were carefully posed *aide-mémoire* of particular things, or people, or scenes. When people wrote about what made these photographs a special form of external memory, what they had in mind was their visual fidelity and fineness of detail. These photographs were not special for being taken at a particular time, and indeed, they represented things that did not change quickly. (Early photography required long exposures, precluding much movement.) By the early 1960s, exposures were much shorter, and cameras were cheap and portable. Although still photography had existed for well over a century, and the dominant corporation in the field, Kodak, had energetically promoted the idea of the "snapshot" for half that time, snapshots looked different now. Not least, they existed amid a world filled with moving images. By the 1960s the cinema, home movies, and television had redefined the status of image technology in relation to memory. Motion and the passage of time had become fundamental to the role of cameras in preserving records of family life.

Kodak reinvented photography and, arguably, memory in the early twentieth century by mounting a campaign to make amateur photography into a means of archiving family experiences, and to create in every American family the imperative to assemble such an archive. The ease of recording personal experience in a "snapshot" was pitched as more valuable than the ability to produce a photograph with specific detail. Photo albums became chronicles of a family's "private story," exchanging the old staged formal images for snapshots of family members in action. The campaign was rather protracted, as technological change and marketing strategies intersected. Kodak first promoted the Brownie camera as a way of preserving important family "stories," "hours," or "times"—it was a way of evoking a period of life or an extended experience. In the postwar years the "hours" shrank to "moments." The change is significant not merely because the unit is smaller, but because "stories" or "hours" suggest the character of life lived at a particular time. When in the postwar years the advertisements began to pitch "moments," they gave the impression of a freeze-frame of a life spent in motion.

In the 1950s and 1960s, snapshots thus came to be identified with the ascendancy of the "moment"—a particular split second caught in the flow of experience—as an entity and a goal, both in photography and in memory. The most influential expression of this idea was advanced by Henri Cartier-Bresson, whose 1952 book on photography, *The Decisive Moment*, had an enormous impact on how still photography was understood. "In photography, the smallest thing can be a great subject," Cartier-Bresson maintained. "The little human detail can become a leitmotif." This "human detail" might be a climactic instant, such as a key move in a sporting engagement, or a moment of critical emotional importance. Cartier-Bresson saw his own strategy as that of achieving, simultaneously and in a fraction of a second, the recognition of "the significance of an event as well as of a precise organization of forms."[10] Cartier-Bresson further characterized the role of the photographic reporter by declaring, "I was there and this is how life appeared to me at that moment."[11] When they were published or otherwise distributed to large readerships and viewing audiences, the highly personalized "moments" captured by photographic reports eventually became part of the collective memory of an event.

The rise of home movies supported the idea that life in motion was something that could and must be preserved for future use. Kodak's advertising

Keep a Kodak story of the children

Autographic Kodaks *$6.50* up

Eastman Kodak Company, Rochester, N. Y., *The Kodak City*

Advertisement marketing Kodak film as a way to capture the essence of family life—its "stories" or "hours" (ca. 1920).

Advertisement presenting Kodak film as a way to capture special "moments" (ca. 1970).

campaign made archiving personal life into a domestic duty, alongside bringing up children to be polite or feeding them healthful foods. The photographic "moment" could be considered a postmovie trend, in that the still photograph could now be seen as a highly specific fraction of a second, frozen in time and separated from a series. In the case of the Zapruder film, of course, what people saw and described as part of their own memories was not the film itself but frozen *moments* captured from it in stills. JFK conspiracy theorists analyzed the film frame by frame, becoming so conversant with salient frames that they referred to them fluently by number. These pictures were tied to past experience through their specificity—the fact that they *were*

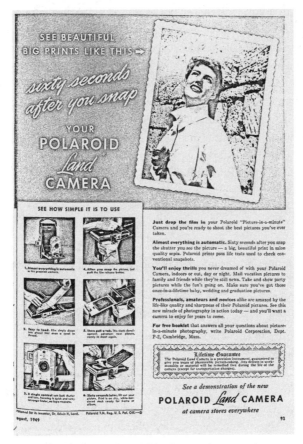

Polaroid advertises immediate memories—"lifelike" images that are ready in sixty seconds, to be enjoyed "while the fun's going on" (1949).

moments—in contrast to the earlier period, in which photographs evoked the feeling of poses and periods of "hours."

The technology that epitomized this new convention was the Polaroid camera. Edwin Land marketed his first instant camera in 1948 as an immediate way of turning experiences into memories, and of making remembering an experience part of the experience itself. Polaroid's advertising was in one respect similar to the trajectory of Kodak's. Soon after it was introduced, the Polaroid camera was described as producing pictures "in a minute"; from about 1960, the "minute" became a "moment."[12]

A commercial for Polaroid in the 1960s, "Take nothing but Polaroids," suggested that this camera could capture instants of enjoyment, enriching an experience by adding instant memories to it without the distraction of fussy procedures.[13] Polaroid pictures were not good archival documents in terms of preservation: the colors faded over time. Their appeal lay in creating an external memory of an event as it occurred, without disrupting the event itself, so that that memory became part of the original experience. In the 1970s and early 1980s, Polaroid could brag that it was the "widest-selling camera in history,"[14] thanks to a mature, fully black-box version that produced pictures in a simple way.[15] With it came the idea of an instant "print" of experience. That notion not only had popular currency but was chewed over by academic heavyweights such as Roland Barthes and Stephen Heath.[16]

Now Print!

One theory about memory of the 1960s was articulated in particularly "photographic" terms and became part of the explanation of flashbulb memories. In 1966, MIT neurophysiologist Robert B. Livingston announced a theory of memory that he termed "Now print!"[17]

Livingston pursued a broad range of research. He had never devoted himself specifically to the study of memory, but among his interests were the neurophysiology of conditioning and reinforcement, and particularly the brain mechanisms of conditioning.[18] He theorized that experiences of unusual importance to an individual were stored in the brain in a special way: a "discharge [was] distributed throughout both hemispheres," which communicated a "print" order for memory, whereupon "all recent brain events, all recent conduction activities will be 'printed' to facilitate repetition of similar conduction patterns." This was a special case of the kind of neural facilitation that he understood to be involved in ordinary memory storage. There were all kinds of stimuli and events in the brain that were not remembered: "Only those occurrences that are biologically meaningful, only those that receive *reinforcement* by a sensory reinforcing signal, like pain, or a sensory approach signal like a 'bug detector,' or by a central reinforcement signal can contribute to the third, fourth, and fifth stages of this postulated 'Now print!' sequence."

The Kennedy assassination was the perfect example. All readers, Livingston thought, would have a "Now print!" memory of it: "You can probably tell us where you were, with whom, and very likely whether you were sitting, standing, or walking—almost which foot was forward when your awareness became manifest. Some people can even tell the details of the dials of a radio they happened to hear the news from, and so forth."[19]

When an individual experienced an event of great significance, the brain "printed" an "associated remembrance" of everything, regardless of whether it had "any real significance for the central matter involved." The cause was a neurohormonal influence that favored "future repetitions of the same neural activities." The growth might involve first an "alteration of ionic barriers," then a replication of proteins, and finally a growth of synapses according to "particular local configurations that will increase the likelihood of repetition of the patterns of impulse conduction that just transpired."[20] Livingston's term reminded some neuroscientists of the instantaneous picture produced by a Polaroid camera, and they often described the event as creating a "photographic image" of a certain moment in the moving pattern of neural activity.[21]

Livingston promoted his idea in a series of neurosciences conferences, where it found common ground with many contributions to a synaptic account of learning. One psychologist who was excited by this concept was Neal E. Miller of Yale, now remembered for helping to develop biofeedback theories. These theories suggested (then very controversially) that the autonomic nervous system was capable of learning, just like the voluntary nervous system.[22] Miller sought an explanation for the process of reinforcement that was based on neural network theories. He filmed a rat on an electrified grid, which finally managed to rotate a wheel that turned off the shock. Initially the act was "not vigorous," but then, "almost like an actor doing a 'second take,' he starts to rotate the wheel furiously." Miller speculated that the sudden relief from pain "produces an automatic increase in the activity of any neural circuits that have just been firing [and that] it is this energization that is responsible for the strong performance (in this case appearing as vigorous overt rehearsal) which in turn is responsible for learning by contiguity." He hypothesized what he called a "go" or "activating" mechanism in the brain that would "intensify ongoing responses to cues and the traces of immediately preceding activities." The stronger the activation of this "go mechanism," the stronger the intensifying of the cues. The parallel to the "Now print!" idea seemed obvious and appealing.[23]

The "go" and "print" ideas seemed to intimate the beginning of an explanation for how the brain could distinguish particularly significant events.[24] Discrimination was a prerequisite for remembering anything, because it was a basic means of noting a change in perception. But at the extreme it produced the kinds of memories Livingston was describing in his "Now print!" theory. The "go" mechanism might also explain why "Now print!" commands triggered the "printing" of *everything* an individual experienced, "even aspects which themselves are not pertinent to the meaningfulness of the occasion." Everything obtaining at the time would be subject to the same instruction, which took the form of "a growth stimulus or a neurohormonal influence that will favor future repetitions of the same neural activities."

The Human Archive

The Kennedy assassination was not just a convenient example for Livingston. He had a professional interest in the case itself. At the time of the assassination, Livingston was scientific director of both the National Institute of Mental Health and the National Institute of Neurological Diseases and Blindness. Partly because of his position, and partly because he had expertise "pertinent to the conduct of the President's autopsy and interpretations of damage to his nervous system," he had followed the news especially carefully in the hours and days after the assassination. When he learned about a small bullet hole in Kennedy's throat, which he suspected was an entrance wound (and if so, evidence of two gunmen), he called James Humes, the doctor charged with Kennedy's autopsy, and warned him that this bullet hole would require special scrutiny. During the conversation, Livingston later reported, Humes was called away from the phone. When he returned, he said the FBI had instructed him to end the conversation, and that was the end of Livingston's involvement in the case.

Livingston's memory of this conversation later became relevant. It soon emerged that the autopsy had destroyed the hole before Humes could examine it for evidence of a shot from the front. Humes explained that he had learned of the hole's possible significance only *after* completing the autopsy. Years later, Livingston publicly accounted for the discrepancy by recounting his own memory of the conversation. In a lengthy statement mentioning his "direct, personal experience" of Hume's work during the autopsy, he argued that Hume statements showed that he was under "nonmedical control" during and after the autopsy.[25]

Livingston eventually became a proponent of "conspiracy" arguments about the assassination—that Kennedy had been assassinated by multiple gunmen, and that a small but extremely powerful group within the American government had exerted "massive" control over the case, destroying and fabricating evidence to support a lone assassin theory. Livingston himself claimed that "the Secret Service had ordered from the Ford Motor Company a number of identical Lincoln limousine windshields." The point was that the windshield in the National Archives might have been a fake. They might have been switched to conceal evidence of a gunman shooting from the front and piercing the glass in the upper left-hand corner, an angle that would have been impossible from behind. If all evidence had been filtered through government procedures and spaces, then physical and documentary evidence must be viewed askance. Conspiracy theorists trusted only evidence from outside government control—eyewitnesses and records made by amateurs. It was essential to believe in the reliability of individual memory, because they could trust nothing else. It was even more useful, therefore, to believe that memories of shocking events were *especially* reliable and that "unimportant" contextual details were also unusually well preserved.

The remaining source of reliable information was the testimony of trustworthy individuals—trustworthy because long-standing conventions of civility provided ways of supporting claims of acting in good faith and disinterestedness, and reliable because of prevailing confidence in eyewitness testimony. Livingston did not explicitly invoke the "Now print!" hypothesis in support of the physiological integrity of eyewitness memories. But he developed it during the same period when he was reading eyewitness narratives of the first few seconds of the assassination and doubting the official evidence. The very point of these narratives was that they featured "Now print!" details of the "contextual" kind whose significance was not initially apparent, such as Kennedy's moving his arm as if to adjust his tie, or the precise location of a hole in the windshield of his limousine. On the day of the assassination, no one expected memories of this kind to be important, since it was possible to study Kennedy's body, his car, and so forth. Only when people began to suspect that the official evidence could not be trusted did this information become crucial.

Conspiracy theories of the assassination simmered through the 1970s and 1980s, occasionally boiling over in new books about the assassination. Livingston himself does not seem to have written publicly about them until the end of his career, in 1998, when he contributed to a book called *Assassination Science*. But his "Now print!" thesis was well known in the 1970s and was consistently associated with memories of the Kennedy assassination, which had after all been Livingston's own key example. So it was that in 1976 Brown and Kulik remembered Livingston's argument and seized on it. They had documented that an event had to be both novel and consequential in order to result in a flashbulb memory. These preconditions seemed to match Livingston's description of the "Now print!" process.

But what purpose did Brown and Kulik think this function served? One salient feature of the Polaroid was uniqueness: a one-of-a-kind object, tied forever to the moment of experience. Unlike other forms of photography, there was no negative to make copies from. The record was unique. Brown and Kulik made a similar argument about flashbulb memories. The idea of "printing" could call up the idea of publication, which reminded them that there was really no need for one brain to remember so well, since external records would do that for us. But before the time of record keeping, such detailed memories of important events would have been important. The "Now print!" mechanism, they thought, was a physiological remnant of a past long before historical records, in the epoch when memory was the only means of preserving knowledge for future generations.

> To survive and leave progeny, the individual human had to keep his expectations of significant events up to date and close to reality. A marked departure from the ordinary in a consequential domain would leave him unprepared to respond adequately and endanger his survival. The "Now print!" mechanism must have evolved because of the selection value of permanently retaining biologically crucial, but unexpected events. It seems to be an irony of evolution that it is just the central newsworthy events that no longer need to be retained because cultural devices have taken over the job.[26]

It was ironic that a neurological device from the mists of the evolutionary past should operate like a modern technology—like a printing machine or, as it was put more commonly, a camera.

Rebuttal and Revision

The concept of a "flashbulb memory" caught on, and not only within academic psychology. Only Gary Larson could claim that even Bambi's forest friends traded flashbulb memories of the day his mother died, but the term certainly came to evoke a kind of memory that was at once individual and collective.

Every time shocking news of a disaster was announced in the 1980s and 1990s, journalists and researchers collected and documented flashbulb memories that rememberers could observe in themselves, probing the effect to assess the role of surprise, significance, emotion, and "consequentiality" in these memories. They sought to discover whether people's confidence in their memories' reliability and permanence was justified, and whether the number of times they had repeated, or rehearsed, their stories was a factor. One of the basic questions was how far flashbulb memories were fundamentally similar to or different from ordinary memories, in the way they were formed and in the kind and quality of information they contained. So whenever a shocking event made international news, researchers sprang into action, typing up surveys to ask their colleagues and students where they had been, what they had been doing, and what they were thinking when they heard the news.[27]

One of the first of these studies was on the attempted assassination of President Ronald Reagan in March 1981. One month after the event, cognitive psychologist David Pillemer surveyed Wellesley College students about their memories of hearing the news. A second round of surveys went out six months later. Most people reported flashbulb memories, even though they also rated the personal consequence of the event as low; the initial "affective" (emotional) response to the event was correlated with the vividness and stability of the memory. Pillemer concluded, like Brown and Kulik, that a special neurological mechanism was at work, but that it centered on emotional arousal rather than "consequentiality."[28]

Yet not everyone thought that a distinct mechanism was at work, or that flashbulb memories were necessarily special in any way other than in the confidence rememberers reposed in them.[29] The first loud note of skepticism

More facts of nature: All forest animals, to this very day, remember exactly where they were and what they were doing when they heard that Bambi's mother had been shot.

was sounded by the eminent cognitive psychologist Ulric Neisser, one of the founders of the field of cognitive psychology. Neisser thought Brown and Kulik had moved too quickly from an observation that people had confidence in their memories to a conclusion that the memories were indeed unusually reliable. People *thought* these memories were unusually vivid and detailed, and they felt sure the details were true. But Brown and Kulik had not tested these claims, and there could be other explanations for people's belief in them besides their veracity. Neisser recounted his own recollections of the bombing of Pearl Harbor. He had vivid memories of listening to a baseball game on the radio that morning. But he checked sports records and realized this was impossible: there had been no baseball game that day. He concluded that most likely it had been a football game. The significance of the contradiction was that his memory was strong but wrong.[30]

Neisser argued that because memories of shocking public events were more likely to be discussed and repeated several times, these rehearsals were likely to intensify one's confidence in memories, whether or not this confidence was warranted. He also had an explanation for why contextual information would feel significant to a rememberer, regardless of whether this information was in fact particularly well retained.[31] Normally, he wrote, "we live in a world of two pasts—the thin string of events in our personal lives and that more grand succession of events that defines the course of human affairs. Certain dramatic but relatively rare events fuse these two segregated sequences into a single stream of remembrances." Hence, "we remember the details of a flashbulb occasion because those details are the links between our own history and 'History.'" Neisser concluded that memories of those events were not "momentary snapshots." They neither had a photographic quality (in the sense of preserving detail) nor referred to a highly specific moment. Indeed, people often combined details from an extended period of experience into this memory that seemed to belong to a few seconds. So flashbulb memories should instead be seen as "enduring benchmarks": the "places where we line up our own lives with the course of history itself and say 'I was there.'"[32]

The chance to carry out more rigorous tests came in 1986, as the debate over flashbulb memories was becoming more intense. On January 28 the *Challenger* space shuttle exploded. While newspapers were consumed with the story and their colleagues were distracted by shock, a few psychologists moved fast to begin a new round of experiments. It was a perfect opportunity: widely publicized and with the kind of shocking emotional impact that people had previously looked to for triggers of flashback memories. A group of psychologists at Johns Hopkins were able to begin their study three days after the event. They were after evidence for or against the idea of a special mechanism, since this was the perfect set of "triggering conditions for the special flashbulb-memory mechanism."[33] They asked detailed questions about respondents' memories, then repeated the questions nine months later to see how memories might have changed over time. They found that over time memories became less detailed, and sometimes inaccurate, and that claims were introduced that had not been there in the original narratives. Their conclusion was that although people were unusually confident in these memories, they were not more reliable than ordinary memories. There was no special mechanism at work.

Ulric Neisser and his colleagues at Cornell also made a study—they had managed to get their surveys ready within hours of the disaster, so that students could fill them out in class the next day. Then they waited two and a half to three years, until late 1988–89, to ask all the students they could find to take another test.[34] They too found that flashbulb memories were far from error-free, although this study detected little relation between emotional response and strength of memory. These memories were not markedly (if at all) more reliable than ordinary ones.

Michael McCloskey, Neisser, and others at length developed an alternative to the photographic model initially proposed by Brown and Kulik. These memories were not like photographs because they did not record events perfectly or permanently; a complex tangle of factors affected them, not just surprise and consequentiality. The account of flashbulb memories that was emerging in the 1980s and early 1990s was sometimes called the "comprehensive model," because it brought together affect, knowledge, interest, and importance rather than just consequentiality and surprise. By now flashbulb memories stood among a cluster of highly contentious psychological effects. Complicating their evaluation, several frameworks that had previously given them meaning—the notion of a traumatic memory, the culture of photography, and the neurosciences of memory—were all shifting dramatically.

The Fading of Flashbulb Memories

In the 1990s the long era of the film snapshot began to wane. People still talked about "Kodak moments" and bought cameras to preserve family events and personal experiences, but increasingly these cameras were digital. The significance of the digital photograph was inevitably likely to be very different from that of the film snapshot. In the film era one would try to time the "snap" of the shot carefully, to pick the best moment from the infinite choices available. In early digital photography this was impossible, at least with the cheaper cameras that were the twenty-first-century equivalent of the Kodak Brownie: the new cameras were too slow to react. After the shutter control was pressed, there was a slight lag before the picture was taken. This "shutter lag" ranged from three-tenths of a second to well over a second, which meant it was nearly impossible to time the shot to capture a particular moment. Instead, people often took a rapid series of pictures, fishing for a good one. Moreover, there was no cost to doing this. Back in the days of film cameras, choosing one's moment was important partly because one paid both for the undeveloped film and for developing. A series of bad shots was a waste of money. Now bad pictures cost nothing, because they could be deleted from memory before any were printed. The identification of shooting with preserving a valued moment dissolved.

At the same time, research in psychology revisited the principle of the enduring, veridical trace. Cognitive and social psychologists had long held to a reconstructive account of memory, according to which memories were not static, permanent "imprintings" of past experience but evolving entities that changed in response to new experiences. Such views dated back at least to Frederic Bartlett. In the contexts of forensic hypnosis and flashbulb memories they were less visible, but in the 1990s they returned to prominence amid criticism of the radical uses that imprinting conceptions were being put to. Reconstructive accounts claimed a revived credibility and political urgency and, consequently, a presence in public culture.

One group that seized critically on the idea of "flashbulb" memories was an interdisciplinary community of researchers in the neurosciences who were interested in consolidation.[35] This term refers to processes that stabilize memory records, both in the first hours after an experience and later, over months or years. The idea that memories become *more* secure some time after an experience is in fact an ancient one, and it has been the focus of research and reflection by psychologists since the beginning of academic psychology in the late nineteenth century. But the topic took on a new salience in studies of retrograde amnesia, with examples of people who could form short-term memories but could not turn them into long-term memory. In the 1970s and 1980s psychologists sought to determine when and how memories were particularly malleable; forensic psychologist Elizabeth Loftus would base much of her career on demonstrating how memories of a particular event could be altered when individuals were exposed to new information shortly after the event, or when they had reason to recall their memories. And initiatives to develop animal models of consolidation opened up new ways to search for a neurophysiological explanation of these processes. Researchers interested in the relation between memory and emotion—particularly anxiety and fear—

became interested in the idea that strong emotional responses at the time a memory is encoded produce hormonal-chemical conditions that lead not quite to "Now print!" instructions, but to special consolidation events. In particular, James McGaugh, a neuropsychologist who made his name in the 1960s and 1970s studying consolidation and amnesia,[36] became interested in the role of emotion in actively preserving vividness and detail in long-term memories.

The perspective on flashbulb memories developed through work on the neurosciences of the emotions represented something of a return to the Brown and Kulik thesis. Beginning in the late 1990s, there was a call for an "integrative" approach to these memories, a compromise between the original claims of the 1970s, and the "comprehensive" model that denied the special power of emotional response, and the specially reliable character of the resulting memory. This view advanced the idea that the initial trigger for the development of a flashbulb memory is surprise associated with the event, which in turn triggers an emotional response if the event is consequential. The emotion is key to the creation of a flashbulb memory, because it strengthens the association, thereby enabling the vivid recall of the event that is characteristic of such memories.

From Shocking News to Repressed Memories

The idea of unusually vivid, and unusually reliable, memories proved appealing in another context, where they came to be directly involved in disputes about how to interpret memories more generally. The vividness and apparently perfect preservation of detail that characterized "flashbulbs" led many to group them in the same category as the then emerging concept of post-traumatic stress disorder (PTSD).[37] An individual suffering from PTSD was unable to form normal memories of the original traumatic event. This could result in a number of symptoms that might seem, on their face, irreconcilable: victims might seem to have no memory for the traumatic experiences, or they might suffer recurrent, uncontrollable flashbacks of the event. In other words, they might remember either too poorly or too well.

The cold war forensic projects of the 1950s and 1960s might seem to have had little in common with the world of trauma research that developed in the 1980s and 1990s, but in fact the extraordinary successes of the forensic hypnotists produced developments and stories that lasted and that launched the careers of key players in the memory wars. Psychotherapist Lenore Terr, for example, spent much of her career developing arguments about trauma, some of the earliest being drawn from the Chowchilla kidnapping of 1976. She interviewed some of the children involved in this case and concluded that single or brief traumatic events were likely to produce flashbulb memories—and therefore to be remembered "extra well." She also came to argue (based on later work) that chronic traumatic experiences like childhood sexual abuse might be blocked from consciousness, producing the psychogenic amnesia that survivors sometimes reported.[38] Terr's argument was influential in the 1980s and 1990s, and it came to underpin popular psychotherapies.[39]

Terr drew consistent skepticism from cognitive psychologists, even as claims like hers gained in popularity with psychotherapists. In 1996 Daniel Shachter, an eminent cognitive psychologist at Harvard and the prolific author not only of academic publications but of popular books on memory, responded powerfully to Terr's argument about flashbulb memories and PTSD:

> Hundreds of studies have shown that repetition of information leads to improved memory, not loss of memory, for that information. To produce profound amnesia, the repression mechanism would have to be so effective as to succeed despite the normal tendency for repeated experiences to enhance memory. People who live through repeated traumas in war generally remember these terrifying experiences all too well. An individual experience or trauma may be set aside, especially when much time has passed, but with rare exceptions [in] such as fugue states—which are generally of short duration—people do not forget an entire set of repeated traumas.[40]

The concept of the flashbulb memory supplied certain key ingredients to the emerging concept of "recovered memory" in the 1980s. The central claim was that traumatic memories were associated with reliability—and, moreover, with unusual fidelity. Recovered memory theorists sometimes borrowed from flashbulb memories this link between fidelity and trauma and attached it to a new kind of memory in which the event was not continuously remembered and reiterated as in the case of "flashbulbs" but was thought to lie dormant for years. Beginning in the late 1980s, flashbulb memories were linked to PTSD in a manner reminiscent of Lenore Terr's argument.[41] Psychological writers began to find flashbulb memories in survivors of all kinds of traumatic events and linked their "flashbulb" character to posttraumatic stress symptoms. Terr's claims became part of a widespread emphasis among some psychotherapists on the idea that traumatic memories were in some sense a "literal" form of the original experiences, preserved in raw form because they could not be transformed into representational memories.

Bessel Van der Kolk and Cathy Caruth are good examples of trauma writers who, in their different ways, understood traumatic events as being intact events preserved neurophysiologically in the brain. As such, these traces ought to have a special evidential and epistemic status that ordinary memories—or perhaps one should say, any memories—did not have. Van der Kolk, for example, did not flatly assert that it was possible for the brain to "take pictures" or for raw sensations to be "etched onto the mind and remain unaltered by subsequent experience and by the passage of time"; but as Ruth Leys points out, his project rested on the assumption that this was so.[42]

This interest in "literal" memories among trauma researchers, who embraced flashbulb-type memories as part of an assortment of traumatic memory records, did not mean the psychological community that had been the original audience for the Brown and Kulik paper accepted the concept uncritically. The idea that one-off traumatic experiences could produce unusually faithful records was not a matter of public controversy when Brown and Kulik first published. In the late 1980s it was, because it had come to be a central issue in the "memory wars."

8

THE LAW OF MEMORY

In January 1989 Eileen Lipsker, a California housewife, reported a memory experienced as a kind of flashback. It had all the vividness and photographic feel of a flashbulb memory, but with a difference: this was the first time she was able to remember the shocking event from her childhood. As Lipsker's seven-year-old strawberry blond daughter sat drawing in a sunny spot on the living room floor, she suddenly looked up and twisted her head to look at her mother. Then came the memory.

As psychiatrist Lenore Terr described it, Lipsker "remembered it as a picture. She could see her redheaded friend Susan Nason looking up, twisting her head and trying to catch her eye." Lipsker herself was standing "outdoors, on a spot a little above the place where her best friend was sitting." Then she "felt" something move off to one side, and she was aware of the silhouette of her father. His hands were raised above his head, holding a rock: "He steadied himself to bring it down. His target was Susan."

This twenty-year-old memory of the murder of Lipsker's best friend when she was eight had apparently been triggered by the homology between the image of her daughter—the reddish blond hair, the movement of her body as she turned to look at Lipsker, the angle of her upturned face—and the fine texture of the "picture" she remembered, innocuous in itself but linked to an event so horrible that, Terr claimed, Lipsker could not bear to allow it to remain in her consciousness.

This story began one of the most sensational murder trials of the 1980s—the trial and eventual conviction of George Franklin for the murder of Susan Nason, almost exclusively on Lipsker's testimony. The memory, however shocking, took much of its plausibility from what commentators described as an almost photographic or "cinematic" character.[1] But this was no flashbulb memory. It was a particularly shocking, and legally consequential, example of a trend that was sweeping the country in the 1980s—recovered memories of childhood events so terrible that they were barred from consciousness.

The 1980s saw a sudden wave of dramatic, terrible revelations about acts of mind-numbing destructiveness—rape and other violence propagated against children by their own parents. It was just as shocking that these claims were being made decades later, by adults who explained that they had recalled the abuse only after years of traumatic amnesia. Lipsker herself claimed to have recovered memories not only of the murder of her friend, but of rape—of the friend, and on several occasions, of herself too. The horror of these crimes intensified the urgency of bringing the perpetrators to justice. But the very reason justice was so desperately needed—the horrible toll these crimes took on their victims—also made legal action almost impossible. By the time the victims were psychologically strong enough to remember and report what had happened, the statute of limitations had run its course. State legislators were appalled that such extreme crimes could in effect create their own legal immunity. Finding it impossible to change the law to make it easier to address them in criminal court, they sought to reform civil law: they revised the statutes of limitations to allow victims to sue their attackers many years after the event.

Remembering Incest in Court

Psychological theories that have been enshrined in the law have a special kind of cultural heft, a legitimacy that is institutionalized—and that also explains itself, because legislative sessions are recorded and transcribed for posterity and judges often justify their decisions in written opinions. These explanations provide the materials for an account of how psychological claims become meaningful and authoritative beyond the writings of psychologists.

The story of how these arguments came into the courts begins a few years before the first suits were filed. The feminist movement of the 1970s made incest a central cause. American society suddenly "discovered" childhood sexual abuse during these years, and feminists quickly came to see it as a symbol of the wrongs of a patriarchal social order.[2] An introspective turn within the feminist movement encouraged autobiographical reflection, and it also led to feminist psychotherapies that sketched out a trajectory from a dysfunctional childhood to a troubled adulthood, and to an eventual triumph brought about by a sustained and carefully managed project of self-study.

This kind of reflection was part of a broader culture of intensive looking back. "Repressed memory" and "recovered memory" entered the public vocabulary in this period, as a variant of posttraumatic stress disorder (PTSD)—itself codified in 1980 after becoming a rallying point for veterans of the Vietnam War. There was also a rapidly growing list of new techniques for eliciting testimony and memory when they were not readily forthcoming. There were conversational approaches that promised to coax stories from traumatized children or battered women, props like pictures and anatomically correct dolls, and the psychoactive forensic techniques discussed in chapter 6. These practices were all controversial. Indeed, some of them were at the heart of a notorious series of accusations of sexual abuse in day-care centers that seemed to point to rampant abuse within American society, only to be sweepingly discredited years later.

In the early 1980s, significant numbers of adults began to complain that they had been abused as children. A few isolated cases soon became a stream, and the trend developed its own momentum. Prestigious therapists like Judith Herman of Harvard and survivor celebrities Ellen Bass and Laura Davis made the rounds of the lecture circuits; support groups, twelve-step programs, and survivor networks sprang up.

Writers on incest used the concept of PTSD to explain the repression of memories of abuse. To survive, they maintained, victims' conscious minds rejected the information, which was then stored in a dissociated form. Even when a victim remembered the events, she often denied their emotional freight. Feminist analysts wrote that no one could "clarify his situation" until he could "feel" the memories rather than "discuss them."[3] As adults, victims suffered from a host of problems: eating disorders, sexual problems, substance abuse, and depression. Even if they remembered the abuse, survivors were unlikely to understand its effects on them until they sought therapy as adults.

One might object that these problems are common. How could one tell whether they resulted from the abuse or from some other cause? Surely some of these diagnoses were straws grasped by troubled people searching for someone to blame besides themselves. But defenders of the diagnoses claimed it was unusual to have several of these problems at once. And one could, they argued, see the very prevalence of some—like bulimia—as a sign that abuse was far more common than anyone realized.

This emerging culture of survivorship placed self-expression at the center of personal recovery from trauma and incorporated a particular arena for that expression—the legal system—into a therapeutic trajectory. The goal was to allow oneself to articulate the full horror of what happened. Doing this was the central task in a project to rebuild, reform, or "rework" one's self. The smallest, most intensely private setting for this "reworking" was a personal diary or private psychotherapy session. Next there was the somewhat more open, but still intimate and sympathetic, setting of group therapy and survivor groups. These were used, instead of the family, to further elaborate the woman's new sense of herself. Then came communication with the perpetrator and the rest of the family through "letters of confrontation" or formal family meetings. The most public setting was the courtroom, where survivors could engage the wider community, inviting them to judge the abuser and to affirm the truth of what had happened.

But this process was blocked for most survivors, because by the time they spoke up it was usually too late to go to court. Most statutes of limitations barred prosecution or civil suit after a lapse of three to five years, or else one year after the victim reached adulthood. Rape was no exception. Limitations were justified by worries about stale evidence, as well as by the idea that only perpetrators of the most egregious crimes should have to endure a lifelong fear of prosecution. But there were some exceptions to this rule. For instance, in cases like asbestosis, where harmful effects would not be discovered until years after the events that caused them, a "delayed discovery" exception applied. In such cases one needed to justify why the evidence could still be trusted—the classic example taught to law students was that of a surgical sponge left in the patient's abdomen after the operation. No matter how much time elapsed, the sponge would be unambiguous evidence of the surgeon's mistake. If these requirements were met, the statute of limitations "clock" would begin to tick only at the discovery, not at the original moment of harm. Believers in repressed traumatic memory thought this exception should apply: not only were victims initially unaware of their injuries, but their ignorance was one of their symptoms.

Several feminist and civil rights attorneys set out to change the law. The particular case that launched the fight seemed doomed from the start, even though there was no ambiguity about what had happened. Lorey Newlander's adoptive father, Ben Ami Newlander, admitted to having sexual relations with her beginning when she was eleven and continuing until she was fifteen. Newlander had even boasted to several people—and affirmed under oath—that the sex with his stepdaughter was "the best I ever had."[4] Lorey Newlander had a clear memory of the incest but had never reported it because she had thought the pain of these years was in the past. But in 1980, in therapy for other problems, she became convinced that these were caused by the incest. Shortly before her twenty-first birthday she brought suit, but the judge dismissed the case because the statute of limitations had run out when she turned nineteen.[5]

Susan McGrievy, a California civil rights lawyer working for the American Civil Liberties Union, became intrigued by the predicament of incest survivors.[6] When Newlander's suit was thrown out, McGrievy and the ACLU joined the fray. They argued that the emotional injury from the abuse was "latent and imperceptible" until it was expressed in Newlander's late teens.[7] They argued that the statute of limitations "clock" should begin to run only at the point when Newlander became able to remember and appreciate her injuries.

The Newlanders soon reached a settlement, so their case created no precedent. But it did start a conversation in legal communities. Although feminists and psychologists had taken a great interest in incest since about 1980, there was no legal literature on the subject. Litigators trying to create new precedents had little they could cite in court. As McGrievy traveled the lecture circuit giving seminars on the litigation of incest cases at law schools, she asked law students to go to work on the subject. Melissa Salten, then a law student at Harvard, found herself motivated by McGrievy's appeal. She later remembered that when she began her research, the Harvard Law Library card catalog did not contain a single card with the subject heading "incest."[8] Salten persevered, publishing an article in the *Harvard Women's Law Journal* supporting legal remedies for adult victims of childhood incest. McGrievy cited Salten's paper even before it was published in 1984; it then drew a flood of citations as the benchmark work on a legal hot topic. Advocacy groups brought cases in other states and lobbied legislators for statutory change.[9]

One such advocate was Los Angeles attorney Shari Karney, who had personal reasons for representing abuse survivors. One day, as she cross-examined a man accused of abusing his three-year-old-daughter, she became ill and distraught, shouting at the witness and the judge.[10] After two nights in jail for contempt of court, she sought therapy, where she recalled childhood memories of her own abuse at the hands of family members. She had had no memory of it before the courtroom scene.

As part of her therapy, Karney prepared materials for a hypothetical civil suit against her abusers. The exercise propelled her into the real world of incest litigation. Thus began years of passionate advocacy. The motivation was not merely, or perhaps even primarily, the hope of financial compensation. The suits were a kind of therapy, both for the survivors and for the community. They broke the "conspiracy of silence" around incest. Karney also collaborated with Mary Williams, a San Francisco–area attorney with years of expertise in incest cases, to advocate for legislative reform. Their effort united several legal and rights organizations behind the cause and eventually led to a change in the statute of limitations to recognize repressed memories.

Karney's story inspired *Shattered Trust: The Shari Karney Story*, a made-for-TV movie starring Melissa Gilbert, which itself became part of the campaign for legislative reform in other states. Director Bill Corcoran was given the job of putting the story together. The idea of repressed memory made sense to him because he had seen "that level of blocking" long before, when he had worked in a treatment home for emotionally disturbed children. As the film was being made, he remembers, actors traded stories about memories of abuse—of themselves, a member of their family, or a friend. During production, "they were naked on the set with their emotions." The set took on the feel of a group therapy session, since everyone "had some story to relate." Either the event had been "psychologically repressed" or the rememberer "didn't want to talk about it themselves and rationalized it," Corcoran recalled. "We would be shooting or getting ready and people would be in the corner crying because it was so upsetting."[11] Corcoran and his colleagues hoped the same effect would carry through to viewers, as a kind of collective therapy and consciousness raising.[12]

Corcoran's film had an easier passage into public life than the bill to revise the California statute of limitations. California state assemblyman Johan Klehs, the bill's author, wanted to contribute to a more general effort to give greater protection to abuse victims and more scrutiny to abusers. There was considerable support for such projects, but also some wariness. This was the time of the McMartin day care scandal, in which child care workers were arrested for satanic ritual abuse.[13] The case broke in the fall of 1983, and by the end of that calendar year more than 360 children were said to have been abused by two child care workers. But by the time the case came to trial, warnings were being sounded that it was all too much and that the psychologists and social workers who had solicited these statements were not able to vouch for their reliability. Years later, when juries finally considered the claims, no one was convicted. The McMartin case eventually came to be used as a warning about the dangers of overzealous and suggestive interviewing techniques, especially when applied to vulnerable witnesses.

Legislators' stances represented the intense ambivalence of their constituents, both about sensational child abuse cases like the McMartins and about recovered memory cases. Those who thought the cases were a form of public hysteria felt responsible to show restraint. Others thought they were seeing a long-unrecognized epidemic and wanted to reform the law to allow greater vigilance. Kleh's bill proposed that victims could sue up to three years after they first "discovered" their injury. "Discovery" could mean a full case of recovered memory after complete amnesia, or it could refer to cases like that of Newlander, who had always remembered the abuse but only belatedly connected it to her problems as an adult.

The bill came up for discussion, but it died in hearings in spring 1984 because the chair, Elihu Harris, was worried about false cases. Despite powerful, and sometimes tearful, evidence given by supporting witnesses, Harris was more concerned about "revenge-type accusations being made against people 20 years later."[14]

Kleh tried again in the next legislative session, and with some significant compromises and a great deal of publicity, the bill passed.[15] A year later an appeals court further weakened the new statute by refusing to apply delayed discovery to a suit in which the plaintiff had never forgotten the abuse but claimed to have connected it to her emotional problems only when she was an adult.[16] But she testified that she had allowed the abuse to continue because she was afraid to speak out. This fear, the court claimed, started the clock on her cause of action during the period of the abuse. In another case, an adult daughter claimed that her father had sexually abused her during her first five years but that she had repressed the memory until she entered therapy as an adult.[17] This case did not have the same ambiguity about the moment of "discovery," but there was a different problem. The new law *allowed* courts to apply delayed discovery to "sexual molestation" cases,[18] but it did not *require* them to do so. In this case the judge chose not to.[19]

These two cases damped down the prospects of delayed actions in the first state to allow them. But even as they fizzled in California, they exploded into life elsewhere.

Tyson v. Tyson

Nancy Tyson was a twenty-six-year-old resident of Washington State who claimed she had been abused by her father from the ages of three to eleven but had repressed the memories until she sought psychotherapy as an adult. She sued her father in 1983, arguing that delayed discovery should be applied to her newly "discovered" memories. The court rejected the argument, and on appeal the state supreme court decided that the discovery rule could be used only when the evidential risks posed by "stale" claims were trumped by greater risks of unfairness and that in this case they were not.

The Washington court regarded it as axiomatic that evidence grows less trustworthy over time, because witnesses' memories are skewed by more recent experiences. The court had other options for how to represent long-term memories. One influential argument held that newly resurgent memories were *more* reliable than ordinary ones. The court tacitly rebuffed this idea. It was also skeptical that a real memory could become inaccessible and then resurface. The court also differentiated between cases that relied on witness memories and ones where other forms of evidence were available. Tyson offered the supporting testimony of family, friends, teachers, and her psychotherapists, but the court saw the memories of all witnesses as problematic and questioned the therapist's expertise. Psychology and psychiatry were "imprecise disciplines" whose "findings are not based on physically observable evidence."[20]

This sweeping rejection of psychological expertise was contentious. It was based on the distinction made by philosopher and psychoanalyst Donald Spence between "narrative" and "historical" truth. Psychoanalysis produced "narrative" truths, he argued—stories about the past that helped explain the present. These narratives were meaningful because they came from the experience of the patient, but they were not *historically* true in most people's understanding of that term. They presented ideas and concerns that came from the present as if they came from the past. Not only did analysands view the past through present-tinted spectacles, they sometimes created a past that had never been. Analysts knew this but nevertheless tended to treat their clients' narratives as historically reliable. Many of them thought the distinction was irrelevant, and even counterproductive if the goal was to help clients better understand themselves as they were now. But what happened when these stories left the psychiatrist's consulting rooms and traveled to other spaces—the family dinner table, a peer discussion group, or a courtroom?

In psychotherapy, where the patient's experience of the past was the primary concern, it might not matter exactly when, where, and how a parent wronged a child. In court, the reverse was true: only the facts about the original events mattered, not one's present-day perspective on them. Legal rules of evidence were deeply concerned about the risks of making subtle mistakes of memory—mistaking the date of an event, or identifying an innocent party who closely resembled one's attacker. The minutiae of the rules of procedure and evidence were all about minimizing such mistakes. They could not be more unlike therapeutic procedures, which, being directed toward the emotional big picture, validated the subjective vantage point of the client.

Conflating the consulting room and the courtroom had potentially devastating implications, therefore, not only for incest survivors, but also for therapy as an enterprise. Spence argued that since analysts were trained not to make the kinds of distinctions that were crucial in the courtroom, memories that emerged in psychotherapy had no place in court. The Washington court drew from this argument an even stronger conclusion: it lumped all psychologists together, even targeting Martin Orne, whose work had in fact helped lay the foundations for the very skepticism that Spence had expressed.[21] For this court, psychological knowledge and legally viable knowledge were simply two different things.

The Tyson decision blocked recovered memory litigation for a few years, until another case revived it. Patti Barton was an incest survivor who recovered memories of abuse in 1987, when she was thirty-three and undergoing "primal" therapy for panic attacks.[22] Barton's father denied her claims but did, she claimed, promise to pay for psychotherapy. Then he reneged, and she decided to sue. The Tyson decision made this impossible, so she collaborated with attorney Jane Mohr to develop a bill giving adult victims three years to bring suit after discovering abuse. On June 9, 1988, the bill became law,[23] and Barton filed suit that very day, eventually settling the case when her father agreed to compensation.

This news inspired Karney and Williams to mount a new push to strengthen the law in California. They drew up a new bill removing most limitations on incest civil suits, allowing victims to sue up to age twenty-six and even later, with certain procedural constraints.[24] The new bill renewed the old arguments about traumatic memories and the law. Strong resistance came from the California Defense Council, which represented the interests of civil defense attorneys paid by insurance companies, who expected homeowners' and personal liability insurance to be targeted by the new suits. More generally, civil rights advocates worried that the bill would tilt too far against the rights of the accused: memories of long-ago events, repressed or not, were far from ideal cases for the exceptions in the new law. Defenders countered that the passage of time actually reduced problems, because adult victims were more credible, more articulate, and harder to intimidate than children.

Other states soon followed the example of Washington and California. By 1991, twenty-one states allowed exceptions to the statute of limitations in cases of repressed memory, and others were debating new bills. The law in these states varied according to the courts' views of traumatic memory—when it might be trusted and how it could be compromised. In particular, Wisconsin allowed delayed discovery in all incest cases where the victim remembered the abuse long after the events, or only belatedly came to understand its impact. But in Illinois, a federal court allowed only cases in which the victim repressed all memory of abuse; it ruled out cases where victims had always remembered the abuse but only later discovered its effects. These became the two templates for delayed discovery law, with most states adopting the Illinois framework.

Survivor Culture and the Politics of Evidence

If we are to appreciate the significance of this legal reform movement, we must look a little more closely at the support communities that were developing at this time. They took it as a central axiom that articulating what had happened was vital to one's recovery, and that listeners' willingness to believe was an important moral act. This, and the fact that the hardest challenge facing a survivor was bringing herself to recount the truth, meant that skepticism was often held to be tantamount to an attack. Such groups defined themselves in terms of "survivorship." They launched countless personal projects of remembering, which involved particular theories of memory and specific therapeutic practices. They also assumed a mandate to educate others about the central problem that brought them together, through publications and, if necessary, legal action. Support communities were often the first to hear survivors' stories and were trusted to understand them. They played a central part in helping them move to the next, wider circle of publicity, whether by writing letters of confrontation, contacting the press, or taking legal action.

All these practices fell under the broader umbrella of what one might, very loosely, call "truth and reconciliation"—a concept that was becoming widely known in the 1980s, although the South African Truth and Reconciliation Commission would not meet until 1995. Community reconciliations were supposed to facilitate an even more important change, a reconciliation within oneself. Theories of remembering—implicit or explicit—were central to these practices.

As one 1989 pamphlet explained it, "Our minds will protect us now as [they] did when we were younger," by preventing survivors from remembering abuse until they were strong enough to tolerate the knowledge. Once they were strong enough, a chance experience would trigger the recall. At that point, "it may seem that we are seeing a slide projector, with pictures flashing very rapidly before our eyes."[25] The memories did not come back in the usual way because traumatized individuals "dissociated" their memories and often experienced them as if from outside the body, looking on from some distance away. One psychologist told participants at a workshop on ritual abuse that memories of such abuse could "overwhelm" people's ordinary "memory drive" and "shut it off." Records of abuse were then stuck in a primitive, sensory part of the brain that could not "play [them] back" until the individual faced life experiences that resembled them in some small way, and even then the records would not reveal themselves in a straightforward manner.[26]

This explained "body memories," in which one seemed to feel past events in the body, with no sense of hindsight or perspective: "The memories come back just the way they went in as visual images, smells or intense feelings. There's no storyline."[27] It was common to experience "flashbacks," defined as the "presentation of a scene or event that occurred earlier, as shown in fiction or motion pictures." Flashbacks and body memories were a sign of incomplete remembering, so survivors needed to continue to delve for memories until they no longer occurred. At that point, they could feel confident that they fully remembered and were in "control" of the memories.

It was more than a matter of simple control, because remembering offered the possibility for "inner growth" in a more than casual way.[28] Survivors felt that they were, sometimes quite literally, divided selves. Some suffered from multiple personality disorder; more commonly, they understood themselves as having an "inner child" hidden away within an ostensibly adult self. That term has been so much mocked that its serious significance and meaning for those involved are now hard to recapture. Yet it was indeed taken seriously and held to be meaningful. It was promoted by psychotherapist John Bradshaw, who developed a highly popular line of therapeutic workshops, self-help books, and recordings.[29] The "inner child" had apparently been frozen in development by the survivor's earlier experience of abuse. When survivors could acknowledge the past—in support groups, therapy, and legal action—they began to integrate multiple personalities and dissociated memories, thereby nurturing the "inner child." On the road to recovery, they had to make common cause with this archived self who fully experienced and remembered the abuse, and who as a result had never matured. The "secondary problems" associated with abuse—eating disorders, substance abuse, depression, anxiety—would persist until this "little girl" was acknowledged. "For some of us, the flashbacks are a constant reminder that we have not yet finished healing our inner child." The only way to eliminate them was to "believe your little girl when she tells you what happened."

This brought survivors to the importance of straightforward, in some cases unquestioning, belief in personal testimony—both their own and others'. As one survivor put it, "The three most important words to me are not 'I love you' but 'I hear you.'"[30] For others, it would have been "I believe you." Thus the kinds of checks that would be most strongly urged by skeptics as a minimal prophylactic against mistaken beliefs amounted to a betrayal. To honor the inner child, it was necessary to "believe the unbelievable." The child needed to know "how you will comfort her fears and concerns" and stand up for her by accepting the truth of the memories. This was what they had to do, no matter what the personal cost.[31]

The injurious effects of patriarchy formed an underlying theme here. Writers on incest commonly maintained that oppressive and abusive parental authority provided the context and the ingredients for abuse. Such authority even supplied the means of denying the history of that abuse, by pushing it out of consciousness: a victim could not allow herself to remember; a mother could not acknowledge the harm her husband was inflicting on her child; a family would not believe what had happened to one of its members.

Evidentiary claims then became moral ones. An empirical question— Did something happen or not?—was now very likely to turn into a political one: Whose side are you on? And accounts of recovered memory came with political freight: "Believe the child" was the cry of victims' advocates and reformers. Not to do so was to side with perpetrators and maintain the silence that had allowed them to continue their abuse.

Lawsuits were in some ways the ultimate expression of this position. One of the most popular (and, among critics, most notorious) survivor handbooks, *The Courage to Heal*, described litigation as one of the most powerful forms of personal therapy:

> Nearly every client who has undertaken this kind of suit has experienced growth, therapeutic strengthening, and an increased sense of personal power and self-esteem as a result of litigation. . . . [A] lot of my clients also feel a tremendous sense of relief and victory. They get strong by suing. They step out of the fantasy that it didn't happen or that their parents really loved and cared for them in a healthy way. It produces a beneficial separation that can be a rite of passage for the survivor.

Recovering from a traumatic past could only begin with therapy; bringing the process to completion required bringing perpetrators to justice.

Raising the consciousness of the broader community thus went along with raising one's own consciousness. One journal, *The Scream*, urged survivors to "raise your voice" to "scream for those still silent," and by doing this to "weave a web of support" for survivors."[32] Another journal told how speaking out had changed "laws, public perceptions, closed minds and lives."[33] One of the most confrontational groups was an underground network called SOUP (Survivors Oppose Ubiquitous Perps). It provided a service: for the bargain price of ten dollars, subscribers could submit the name and address of a "perpetrator," and this information would be included on a "perp" list that would be circulated to schools, police, and neighbors in the alleged perpetrator's community.[34]

One of the longest-lasting survivor journals was *Write to Heal*, which began in January 1992, and as of early 2010 maintained a healthy paper circulation and a website. Anne Cox, a professional writer, created the journal as a forum for therapeutic self-expression about traumas of all kinds.[35] The title's pun asserted that "healing is an entitlement" of everyone. Most authors were adult women survivors of abuse in their twenties and thirties, but submissions also came from older women, from men, and even from abusers—who wrote about their own troubled childhoods.[36] The healing implied in the title of this journal naturally extended to political action. *Write to Heal* joined the campaign to reform the statute of limitations. Cox wrote to Senator Joseph

Biden, then chairman of the Committee on the Judiciary, urging him to see child abuse as fundamentally no different from other dangers whose effects could be latent for years, like Agent Orange, IUDs, and asbestos. This meant it was crucial to pass legislation mandating that medical offices and schools nationwide "maintain records in electronic storage and retrieval systems," to make it easier to track down indications of abuse after the fact.

By 1990, a national conversation was taking shape around recovered memory. The time was ripe for it, after the steady intensification of state legislation, medical publications, and civil litigation. And cases were multiplying. For instance, Loretta Woodbury was in therapy in Northridge, California, in the late 1980s when she suddenly remembered being abused by her stepfather as a child. At this point she was in her fifties, and one might have thought it was a bit late to take legal action. But she worried that her stepfather might be abusing others, so she alerted authorities in her Wyoming hometown. The local district attorney investigated. He found a wealth of corroborating evidence—including a handwritten plea for help she wrote when she was fourteen and sewed into a relative's coat pocket—and decided to prosecute the seventy-three-year-old suspect. He pleaded guilty on November 28, 1990, two weeks before the verdict in the Franklin case.[37]

But a single event propelled the issue of recovered memory onto the front pages of national newspapers: the George Franklin murder case in which Eileen Franklin-Lipsker accused her father of murder, based on a recovered memory.[38] There were two experts hired by the prosecution, Lenore Terr (the trauma researcher in the Chowchilla case years earlier) and David Spiegel, a clinical and academic psychologist at Stanford. Terr argued that there were two types of traumatic experience, invoking the distinction we encountered in chapter 7. Type 1, a single experience, would be remembered. Type 2 was a pattern of repeated abuse, and this might be repressed. She claimed that though the murder was an extreme event, it was part of a pattern of abuse and therefore fell into type 2.

The detail in Eileen Lipsker's account was one of the features that made it compelling. But although this was how her memory was commonly described, the truth was more complex. The story she related was in fact not so new: it was the scenario as speculated by the police, who (along with Franklin's own wife) had targeted Franklin as a suspect. What *was* new was the claim that Lipsker had witnessed the murder. To her, that details of her refreshed memory had figured long ago in police reports corroborated and confirmed her newly recovered memories. But skeptics pointed out that all the details she related had also been in the newspapers, so they were not really "new" revelations, but the jury did not know this because the judge did not allow the defense to introduce this material. Skeptics also claimed that Lipsker had initially recovered these memories under hypnosis or barbiturates; that it was only when she realized this disqualified her from testifying in court that she claimed the memories came back spontaneously.[39] Testifying for the defense, forensic psychologist Elizabeth Loftus insisted on the reconstructive character of long-term memories in general, and on the lack of academic research supporting the legitimacy of recovered memories.

The jury was now faced with experts on traumatic memories and counterexperts on the unreliability of memory. But since they did not know about the original newspaper reports, they could hardly assess the possibility that they were the basis of the memory claims. They sided with the trauma memory experts and believed Lipsker. On November 30, 1990, they convicted George Franklin, who received a life sentence in the first murder case based on a recovered memory.

With this first murder conviction relying almost wholly on a recovered memory, recovered memory made headlines in newspapers, national magazines, and television news. Something many people had been vaguely aware of was now center stage. New victims appeared and announced themselves— some of them celebrity figures, such as former Miss America Marilyn Van Derbur, who remembered at age twenty-four that her father had abused her when she was a child, and sitcom actress Roseanne Barr, who recovered memories of her own abuse after thirty years of amnesia.

Litigating Repression

By mid-1991 the number of cases of long-forgotten abuse that were being tried in courts across the United States began a steep climb that would not level off for several years. "Recovered memory" was becoming a publicly acknowledged category. It was extraordinary but, almost because of that, plausible. A decade earlier, it had been virtually impossible for an adult to bring suit based on claims of being abused as a child. This was now a potentially viable action in most states—as long as you had the right kind of memory. The best kind to have if one wished to mount a civil suit was not, by ordinary reckoning, a very good memory at all, since one would ordinarily think that the strongest memories would never be forgotten. Here, though, one usually had to have failed to remember the events for most of one's life and to have recalled them only shortly before bringing the suit. Moreover, because of the legal problems that forensic hypnosis had in the late 1970s, the returning memory could not have been brought back by hypnosis.

If one accepted the idea of repressed memory, one could infer more from a belated resurgence of memory than that plaintiffs had had no previous opportunity to sue. One might also claim that the recovered memories were *more* reliable than continuous ones because they had been latent and locked away for the intervening years; this latency implied a kind of protected storage. If one did not believe in repressed memory syndrome, however, this kind of chronology could suggest opportunism at least, and possibly fraud on the part of the plaintiff—a claim of new memories when they were either not new or not real. Either way, the two-category rule that now applied in many states (allowing plaintiffs suffering from repressed memory syndrome and barring those with "good" continuous memories from bringing suit) encouraged claims of repression.

The idea of a sudden abreaction, or a shocking return of memory, presented a kind of all-or-nothing understanding of the return of a lost memory. But in many cases this was not an appropriate representation. Many survivors would have characterized their childhood homes as dysfunctional or even abusive before the moment when they suddenly remembered specific instances of sexual abuse. A survivor might have said, "I always knew something was wrong at some level, but I couldn't face it. Therapy helped me to begin to acknowledge it, and with hard work in therapy I was able to confront my past." The deep amnesia of a "repressed memory" shaded into the softer, more slippery notion of "avoidance," defined here as choosing not to focus on a painful subject. But the law supplied a motivation to present the distinction as categorical.

In this context, abuse survivors—and those who were only beginning to suspect they might have been abused long ago—began to impose new, and changing, categories on their own memories. They not only attended to the content of the memories, but also revised their understanding of what memory itself was and could be. The latter, of course, bore directly on the question of content and fidelity. In so doing, they were responding to a legal culture that encouraged sharper distinctions between continuous and repressed memories, and to a set of therapeutic practices that made it easier to craft them.

This is not to say that those who made strong claims to traumatic amnesia and spontaneous recovery were opportunists pretending to have remembered experiences that they did not in fact have—though doubtless opportunism did sometimes occur. Rather, they understood the range of possibility and plausibility for acts of remembering in ways that had been powerfully affected by the previous decades' intellectual, legal, and social changes. Those changes led to very consequential claims about the past that might not—and perhaps *could* not—otherwise have been made.

For believers, that a recovered memory had been inaccessible was confirmation of its privileged epistemic status: inaccessibility = incorruptibility. For the historical account being presented here, this equation brings us to a climax characterized by a paradox: the acme of memory as recording is a memory that counts as having been recorded faithfully because it cannot be played—until it is, years later. Replaying memories is the equivalent of an archaeologist's digging up an ancient artifact or a hoard of coins, which contains its own truth by virtue of having lain undisturbed through centuries of change.[40] Indeed, discussions of recovered memory by survivors and skeptics alike are awash with archaeological references to digging up "buried" memories or "exhuming" knowledge of the past.

There is a certain purity—and consequently certainty—to archaeological claims of this kind, because the alternative is not to say something different but to say nothing at all. Dug-up relics have no immediately evident context to which one can appeal in interpreting them. The tendency (again paradoxical) is for this to generate exaggerated claims of certainty and importance—another paradoxical effect, since one could argue that the significance of such objects is wildly underdetermined by the lack of context. One might find recovered memory's archaeological associations peculiar, on the grounds that in cases of sustained abuse there ought to be other kinds of evidence to adduce. If this is valid, perhaps what sustains the claim is the accrued authority of the notion of memory as recording. But in many cases there is no other evidence because these memories refer to very private moments from long ago. In this respect, one might see this pivotal episode as a culmination of the "history of private life."

There is a sense in which the legal projects I have discussed in this chapter also represent the ultimate high-water point in the history of the idea of memory as a recording or a personal archive. In the 1980s this idea, represented by the claim of repressed and recoverable memory, finally seemed set to transform the institutions of memory, especially the courts, which embraced it more readily than courts of the 1970s had accepted forensic hypnosis. Scientific and popular writing was representing recovered memories—

underpinned by the idea of a recording—as able to transform the way several generations of Americans thought of their own lives and their families. But a challenge to these claims would begin within the year, organized by Orne and led by Loftus, the last two generations of memory skeptics. The particular arguments they mounted drew not only on their own research—the demand characteristics of Orne and the eyewitness memory problems of Loftus—but also on the revival of a much older view of remembering, the subject of chapter 9.

9

FREDERIC BARTLETT AND THE SOCIAL PSYCHOLOGY OF REMEMBERING

One winter evening in 1974, television viewers saw footage of a man wearing a leather jacket, sneakers, and hat lurking in a doorway. As a woman walked past him down a hall, he jumped out, grabbed her purse, and ran toward the camera.[1] The viewers were asked to choose from a "lineup" of six "suspects"—or to reject them all—and phone in their decisions. Thousands did so, until they swamped the service and the phone line was cut off.

This turned out to be not a real crime captured in progress, but a mass psychology experiment. It recalled the theatrical tricks that Hugo Münsterberg had pulled on unsuspecting law students—but on a much bigger scale. Professor Robert Buckhout of Brooklyn College had arranged the experiment to test the eyewitness memory of television viewers. The "purse snatcher," number two in the lineup, received 302 of the 2,145 votes that came in before the researchers stopped taking calls. This was a pathetic 14.1 percent correct—the same "as if the witnesses were merely guessing," wrote Buckhout in an article pointedly titled "Nearly 2000 Witnesses Can Be Wrong."[2]

Buckhout's experiment was part of a new kind of memory research that was developing in the 1970s. It presented its results as a shocking reality check: memories were not the reliable recordings they were popularly understood to be. Even with the most expert mediation, memories were partial from the beginning, and they could change. It was impossible to tell which parts of a memory were "real" and which were constructions after the fact.

In one sense these claims were indeed new; they were framed in diametric opposition to the recorded memory theories of the forensics and psychotherapy experts. But in a different sense they were not new at all. Researchers saw themselves as extending the work of early twentieth-century psychological researchers whose findings had previously been overlooked, or at least under-appreciated. One of these two ancestors was, as one might expect, Münsterberg himself. Buckhout—and more famous colleagues like Elizabeth Loftus and Martin Orne—saw themselves as finally realizing his ambitions for a forensic psychology. Another ancestor was Frederic Bartlett. Bartlett was a British psychologist based at Cambridge University in the early and middle twentieth century. His work on memory was not widely influential when he first published it in the 1920s and early 1930s, but in the 1970s and 1980s, long after his death, he became one of the most cited psychologists of the twentieth century (if not the most cited).

Bartlett has come to stand for a kind of claim about memory that has become commonplace among academic psychologists: that memories have a constructive or reconstructive character. We do not merely forget information over time, we "reconstruct" the content of our memories by adding to and otherwise altering them. This idea has become central to recent accounts of memory. Bartlett is now claimed to have issued, in the words of one writer, "a consumer alert regarding a common mental product—memory," the point being that memories were not as good, not as useful, not as strong as we would like to believe they are.[3] He stands for a warning that memory could not measure up to the idea of faithful record keeping. So he is now invoked in claims that our memories are "wrong," are "false," or don't serve us well. Thus "Bartlett" has become a useful figure for the developing fields of forensic and legal psychology. But the original Bartlett was quite different.

Bartlett did believe that memories changed dramatically over time, but he did not think this made them false or bad. In fact, he thought they served us better as a result. To understand where the idea of reconstructive memory originally came from and how it changed so dramatically decades later, it is necessary to take a big leap back, to examine the way Bartlett's own work developed in the 1910s and 1920s, a time when the very idea of memory as "faithful record" was also emerging.

One sunny afternoon in May 1913, people lined up outside the newly opened Cambridge Psychological Laboratory to take a tour and try out some experiments. Frederic Bartlett, then a young graduate student, darkened the room. Then he used a special projector to display a series of geometric shapes, pictures, and optical illusions for a fraction of a second each. Afterward, the visitors attempted to draw the shapes they had glimpsed.

The results varied widely. People seemed to bring quite different perspectives to bear on what they were seeing. But as varied as the sketches were, they all tended to be representations *of* something, even when it turned out to be something very different from the original image. Bartlett concluded that the sketches were an attempt to make sense of new information by relating it to preexisting assumptions. He later argued that the viewers had "filled up the gaps" in their memory with "the aid of what [they had] experienced before in similar situations" or what seemed "suitable" to the other information at hand.

What the visitors themselves thought is not known. But for Bartlett this was the beginning of a new account of perception, of memory, and ultimately of thinking itself. In perception the mind actively chose features of the observable world to make meaningful, by placing new information in the context of established ideas. Psychologists of the time (as well as people with no psychological training) often assumed that perception and recollection were quite different things, but Bartlett argued that much of what was assumed to be perception was in fact recollection of very recent events. This claim not only made memory a central issue but also emphasized that it was the active result of thinking and creativity. Convinced of the validity of the contention, Bartlett embarked on almost two decades of research on perception and memory, of which his landmark *Remembering* (1932) was the most significant result. He referred to "remembering," rather than to "memory," because he wanted to underline the active mental processes that he believed made events meaningful.

Bartlett worked in a wing of the new Physiology Laboratory in Cambridge—a world away from the urban American courtrooms, police stations, and medical consulting rooms of truth serum. The forensics pioneers discussed in chapters 1 and 2 all measured memories against an ideal "truth" that individual recollections could in principle conform to or diverge from (Münsterberg thought they must fall short, House that they might be perfect). Bartlett discarded this basic commitment, dismissing as wrong and irrelevant the idea of a permanent memory record and even of "truthful" perceptions. If Bartlett had little to do with the forensic psychologists, neither was he much like American academic psychologists, for all that his laboratory, like American ones, was modeled on German labs. Bartlett's counterparts in United States academic psychology departments largely assumed that, to be studied properly, human behavior had to be simplified, as pioneering memory researcher Hermann Ebbinghaus had done. Ebbinghaus had made subjects memorize nonsense syllables, then charted how long they took to forget them. Bartlett found Ebbinghaus's approach ridiculously artificial because it did not take into account the context in which memories were made and used. Nonsense syllables revealed nothing about how memory worked outside the laboratory, when people were remembering appointments, the names of acquaintances, and events of the distant past.

Bartlett came to believe that mental content had no significance outside the situations that gave it meaning. Rather than treating memories as static records that could be "true" or "false," "remembered" or "forgotten," one should study how both context and meaning changed over time. This required a naturalistic, even an anthropological approach, using techniques that could study how memory functioned in everyday life. His science was a compromise between the laboratory and the field—even if the field, in this case, was the peculiar environment of Cambridge University.

Thinking in the Laboratory

Bartlett's path to Cambridge was unusual. Illness had kept him out of grammar school, so he studied mainly at home. He earned his BA from the University of London by correspondence, then moved to Cambridge, taking a job as a tutor as he sat the moral sciences tripos to prepare himself for postgraduate research. He became interested in anthropology and psychology,[4] and particularly in the work of William H. R. Rivers. There was no graduate program at this time, so Bartlett resolved on postgraduate research under the informal mentorship of Rivers and the other researchers associated with the laboratory. Rivers encouraged Bartlett to take psychology courses as a preliminary to anthropological study, and this brought him into contact with James Ward and Charles Myers (then director of the Cambridge Psychology Laboratory), as well as Cyril Burt (then occupying a temporary position). It also introduced him both to the tradition of British systematic psychology and to its use of the German laboratory psychology of Wilhelm Wundt and Gustav Fechner. He took a position assisting in the psychological laboratory, hoping to produce a research dissertation that could earn him a college fellowship.

Bartlett arrived just before the new, purpose-built laboratory put Cambridge psychology on the international map.[5] He was entering an environment in which researchers were unusually free to define their own questions. The most influential researchers, Myers and Rivers, encouraged an empirical and methodologically catholic climate, one strongly engaged with anthropological issues.[6] Myers is generally regarded as the architect of Cambridge psychology, designing the new psychological laboratory, running the *British Journal of Psychology*, and shaping psychology elsewhere in Britain. But Rivers was even more influential, with broad interests and a huge research output in the 1900s and 1910s that inspired many students.

Rivers began his Cambridge career as a laboratory psychologist, but his interests took a strong anthropological turn after he served as a psychological researcher on the Torres Straits expedition of the 1890s. The expedition had been designed to establish anthropology at Cambridge, using psychology as its entry point.[7] The approaches Rivers and his colleagues developed from these experiences came to permeate what psychology meant at early twentieth-century Cambridge. Rivers became engrossed in psychological and neurological work—on the regeneration of nerves with famed neurologist Henry Head, and on psychoanalysis with traumatized soldiers. When Bartlett came to Cambridge, Rivers was immersed in ethnographic work and was promoting a number of theoretical claims that would deeply affect him.

Myers and Rivers therefore generally steered students away from "brass instrument" work and toward other approaches, often designed to be pursued outside the laboratory entirely. Their followers learned clinical psychology, "abnormal" and developmental psychology, and the psychology of "primitive peoples." Bartlett's Cambridge friends thus drew from many fields and disciplines, encompassing history, mathematics, philosophy, and literature. This environment helped to free him from conventional constraints in the way he framed his research questions—something other psychologists often remarked on.[8]

Bartlett's chief research interest also sprang from this eclectic context. He soon became interested in a process that Cambridge anthropologists were calling "conventionalization." Alfred Court Haddon had used this term to study art in the 1890s, to explain how artistic styles developed from the individual and specific into more abstract and standard forms.[9] Rivers used conventionalization in the 1910s in an argument about Melanesian history: it explained how "a form of artistic expression introduced into a new home becomes modified through the influence of the conventions and long-established techniques of the people among whom the new notions are introduced."[10] As early as 1913, Bartlett himself signed a contract with Cambridge University Press to write a book on conventionalization.[11] The following year, as he began his own experiments, Rivers's book on Melanesia was published and immediately drew controversy for its claims about conventionalization. For Bartlett, the controversy was his chance to show not only that the mind transformed incoming sensations during perception, but also that an individual's social relationships and status mattered too. Rivers's approach linked perception and cognition to social influence and structure and, more generally, provided a framework for making claims about the context-dependent character of knowledge itself.

Bartlett's research was to be carried out under unusual conditions, because just one year after the laboratory opened, World War I began. His mentors left Cambridge: Myers for military service and Rivers for Maghull Hospital near Liverpool. Bartlett's health disqualified him for military service, so he found himself—despite almost no formal psychological training or credentials—in the position of "relief director" of the laboratory. He kept it running throughout the war. Despite his administrative duties and other work, he also managed enough research to earn a fellowship at St John's College.[12]

Reproductions

Bartlett's first experiments used individual screenings of visual images to show that the mind made creative "mistakes" in visual perception. But it was hard to discern clear patterns from single experiments. Ward suggested that Bartlett build the element of time into them systematically, by using "repeated reproduction." Jean Philippe had pioneered this technique in the Laboratory of Physiological Psychology at the Sorbonne. There Philippe had studied research subjects' attempts to reproduce pictures shown to them for brief flashes of time.[13] Bartlett adapted Philippe's techniques for his own use but disliked the distinction Philippe made between perception and memory. Bartlett declined to draw a sharp line between perception, which he argued was often portrayed (as Philippe did) as "fluid," and memories, which were generally assumed (in Philippe's words) to have a "fixed, lifeless" character.

Bartlett adapted the repeated reproduction technique to an experiment suggested to him by Norbert Wiener (who later founded cybernetics). In 1914 Wiener, who visited Cambridge in 1914 and 1915 to study mathematical logic and became good friends with Bartlett, suggested, "Couldn't you do something with 'Russian Scandal,' as we used to call it?"[14] This is a game, also

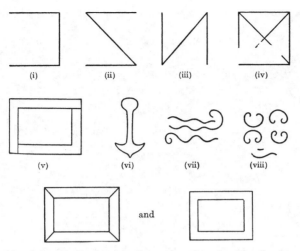

Simple line drawings used by Bartlett in his first studies of visual memory (1916).

called "operator" and "Chinese whispers," among other names, in which a sequence of words is passed from one person to another. After several repetitions it has changed dramatically. Rather than repeated efforts by individual subjects to remember something, the techniques of "repeated" and "serial" reproductions allowed only the first subject to see or hear the original sketch or story; successive subjects were presented with the work of preceding subjects. Both reproduction methods involved a series of attempts to reproduce an image seen only briefly, after either just one exposure, or a series of glimpses (each very short) of the same image, or a series of glimpses of a series of images. All of these were aimed at documenting the ways the mind acts on perceived material to make it meaningful.

Bartlett's first experiments, carried out mainly in 1915 on students and staff at the university, examined differences in how test subjects described various pictures, primarily geometric shapes. Bartlett ascribed the differences he documented to the subjective interests and social orientation of his subjects—a form of conventionalization. He showed subjects these images using lantern slides and a device called a portable Hales tachistoscope, a projector that could display an image on a screen for a precise period, which was varied by the pendulum settings.[15]

Bartlett used this machine to project a series of sets of pictures, each set more complex than the last.[16] His first set, simple stick diagrams, were reproduced quickly and accurately. Next came more complex and more representational

W. F. Yeames's painting *Prince Arthur and Hubert*. Courtesy of Manchester Art Gallery.

sketches. This time people paused briefly before making their sketches. Bartlett also noticed that they always gave a name to what they were drawing. Each name "not only satisfied the observer, but helped to shape his representation." For example, people who described item (v) as a picture frame (see p. 203) drew the right-hand sketch reproduced in the bottom row. In only one case did the reproduced sketch match the original precisely: when the subject described the image as "two carpenter's squares placed together." Bartlett argued that this naming was part of an "effort after meaning." This meant that his work was as much a study of memory (of the recent past) as of perception. People seemed actively (though not consciously) to be *using* memory and imagination as a framework for new information.[17]

Bartlett carried out an enormous number of visual-reproduction experiments of this kind in 1914–16, only a small fraction of which appeared in print. Soon after writing his fellowship dissertation, he carried out more picture drawing research, this time using more complex material—paintings rather than simple drawings. He found that each subject immediately registered some general impression of the picture. After that, successive viewings added details, but only those that fit the first impression. His favorite example involved a painting by W. F. Yeames, based on Shakespeare's *King John*. Hubert de Burgh, commanded to blind John's young nephew, Arthur, listens to Arthur plead, "Will you put out mine eyes—these eyes that never did, nor never did, nor never shall, so much as frown on you?"[18]

People who viewed this painting for a quarter of a second produced strikingly different descriptions. One described a "woman in a white apron with a child standing by her knee. She is sitting down and has her legs crossed." On a second attempt the same person decided the woman was standing up but made few other changes, over a total of thirty-eight viewings. Another observer saw a struggle between two figures. Over the course of fifty-five showings, he produced increasing detail about a "dark fellow" getting "the worst" of a wrestling match.[19]

Bartlett concluded that people were motivated to construct a complete interpretation no matter how little information they took in. Their initial conclusions became entrenched, so that repeated exposure could add detail but could not change this "frame." He named this process "imaging"— the subject created a sketchy account of something, then filled in the gaps. Most people were unaware of this "filling in," but one subject did make some remarks that Bartlett found illuminating. He was troubled as he watched glimpses of a "Margate lifeboat on the slips." After the eighteenth viewing he declared, "It is no use going on. All the time I am getting a suggestion of the docks at home. And they are what I see, not the picture in front of me. One of the first things I did when I got a camera some time ago was to take a picture of that spot at home that I was reminded of. . . . There was a ship of heavy freight there at the time, just as there is in the picture here. So I am always confusing the two, and I shall get no more out of this."[20]

Bartlett also did a series of studies with inkblots, using them in a way that was then standard, as a tool for studying imagination and imaging, with no clinical purpose.[21] (Hermann Rorschach was developing his famous test about this time, and it would later be widely used by psychiatrists, but Bartlett never considered any such thing.) For Bartlett, inkblots were pure raw material, innocent of any inherent representational significance. Whatever subjects made of them would reveal the mind's tendency to construct meaning in particular ways—the general characteristics of "constructiveness," rather than diagnostic information about a specific test subject. Bartlett noted how varied his subjects' responses to the blots were, as well as how "dynamic" (subjects usually described people or animals as doing something) and how complete (using all of the inkblot instead of just one spot). But his overarching conclusion was that the blots needed to be "actively connected with [some other mental content] before they can be assimilated."[22] Bartlett was particularly taken with a comment from one of his subjects, that he needed to "rummage around" in his mind to find something to use to interpret them.[23]

So far, the visual material used in these tests was simple or complex, representational or abstract, but none of it was thought to contain referents to everyday experience. Bartlett now wished to study acts of remembering that were more like tasks of real life. Once again he began with pictures, but these were picture postcards depicting military men.[24] These postcards were appealing because the faces had individual peculiarities, but they were also based on visual conventions associated with certain categories of personnel. He asked his subjects to look at five cards, one at a time, for a set period, then to put them aside and talk about them in the order in which they had been viewed. The subjects were asked to return for a second session two weeks later to discuss the images again.

Bartlett was not surprised to find that the descriptions changed in the second session, but he was intrigued to see that there was a pattern to these differences. Alfred Binet's notion of *idée directrice* seemed appropriate to

Postcard of military "types" used by Bartlett in studies of visual memory, 1921.

one of the most common patterns, the notion that once one seized on a particular idea, this idea would direct the way further information would be interpreted.[25] Bartlett also noted that the parts of the images that changed most over the time between sessions were those that interested the subject most. Bartlett argued that his subjects tried to discern a *rule* "of arrangement" for what they saw. They would then reconstruct their perceptions according to the rule, so as to make the pattern consistent.

So far, it may appear that the mind was a discrete system, taking in information and using existing mental content to construct something meaningful. But Bartlett was further convinced that it was the interaction between people—in broad terms, culture—that most powerfully affected how this process worked. This idea inspired Bartlett's most famous experiments, using a story known as "The War of the Ghosts."

The story was both exotic and filled with narrative disjunctures. It was a translation of a North American folk tale, published in an American ethnology journal. Bartlett was drawn to it by its apparent lack of logic and its strange imagery, which he thought would provide a foil for studying the mind's efforts at fitting new material to existing forms. It would also help him examine the cultural dimension of the construction of meaning—allowing him to explore "what actually happens when a popular story travels about from one social group to another."

The "War of the Ghosts" told about members of a tribe meeting on a river and preparing for war. The story is so critical to Bartlett's work that it is worth quoting at length:

> One night two young men from Egulac went down to the river to hunt for seals, and while they were there it became foggy and calm. Then they heard war-cries, and they thought: "Maybe this is a war-party." They escaped to the shore, and hid behind a log. Now canoes came up, and they heard the noise of paddles, and saw one canoe coming up to them. There were five men in the canoe, and they said: "What do you think? We wish to take you along. We are going up the river to make war on the people." One of the young men said: "I have no arrows." "Arrows are in the canoe," they said. "I will not go along. I might be killed. My relatives do not know where I have gone. But you," he said, turning to the other, "may go with them." So one of the young men went, but the other returned home. And the warriors went up the river to a town on the other side of Kalama. The people came down to the water, and they began to fight, and many were killed. But presently the young man heard one of the warriors say: "Quick, let us go home: that Indian has been hit." Now he thought: "Oh, they are ghosts." He did not feel sick, but they said he had been shot. So the canoes went back to Egulac, and the young man went ashore to his house, and made a fire. And he told everybody and said: "Behold I accompanied the ghosts, and we went to fight. Many of our fellows were killed, and many of those who attacked us were killed. They said I was hit, and I did not feel sick." He told it all, and then he became quiet. When the sun rose he fell down. Something black came out of his mouth. His face became contorted. The people jumped up and cried. He was dead.[26]

Bartlett asked his subjects to read this story and later to reproduce it. Sometimes he asked people to retell it on multiple occasions—this he called "repeated reproduction." In other experiments, using the technique of "serial reproduction," the story passed from subject to subject in the manner of Wiener's "Russian scandal."[27]

One might assume that the story began to change only after the beginning of the experiment, but in fact the text Bartlett gave his subjects was itself the product of multiple reproductions. Bartlett found his text in the *Bureau of American Ethnology Bulletin*, published in 1901 by American anthropologist Franz Boas.[28] Boas had learned of it from his informant Charles Cultee, who gave him many narratives that he had collected as specimens of tales and myths of the Kathlamet people. The Kathlamet (or Cathlamet) were a tribe with their own language, which Boas said was a dialect of Chinook spoken by people who lived far down the Columbia River, including the village of Cathlamet itself, where Cultee had lived for several years. At the time of the transcription, Cultee was one of three living speakers of Kathlamet (the last died in 1930).

Cultee himself had heard the story many times, and he told it to Boas more than once, first in 1891 and then in 1894. There were some differences between these versions, including more explanation the second time around. Both times, Cultee and Boas wrote the story phonetically in Kathlamet and then translated it word-for-word, interpolating English words to preserve the Kathlamet grammar. Boas then edited it to read more naturally in English. The phonetic Kathlamet, the word-for-word English translation, and the more fluent Boas version were all published together. Years later, Bartlett produced his own "slightly altered" version for his experiments, giving the strange story a whole new career in 1920s Cambridge. Bartlett was thus choosing a text that had a kind of reflexive appropriateness to a project about the mutability of knowledge. Its own history (though Bartlett never commented on it) seemed to show that repetitions often remove or add details that help to make sense of the whole.

That this story came from an indigenous people, and was published by Boas, suggests that Bartlett's experiment was connected to an ongoing debate about whether there was any significant difference between the mental capabilities of "primitive" peoples and those of Western, "modern" societies. Boas was a leading figure in the development of American cultural anthropology, promoting an understanding of culture as fluid. He reckoned that individuals' understandings took shape in their immediate social environments, but that they could also change the surrounding culture in unpredictable, locally contingent ways. He recognized no great distinction between the primitive and the modern. Boas's own views on remembering came into play when he criticized an influential account of "primitive" psychology by philosopher Lucien Lévy-Bruhl,[29] who had argued that "primitive" peoples were fundamentally different from "modern" ones: they often had prodigious memories, but they could not reason from them in a logical, analytical way. Boas emphatically rejected this distinction.[30] In the 1920s, Bartlett would develop a similarly toned argument about what would later be termed cognition, though in the 1910s his sources merely hinted at it.

Bartlett's "War of the Ghosts" research was more directly indebted to Rivers's anthropology. Rivers's work in the Torres Straits indicated there was no distinction in perception between "primitive" and "modern" peoples, and he extended this claim in later work.[31] He also began to promote a diffusionist understanding of elements of culture in his researches of the 1910s. If such characteristics could be translated into a psychological experiment—that is, if specific human beings could be shown to process information in a manner similar to the distortions of culture that Rivers had described—this would give powerful support to Rivers's claims about larger groups and societies. Bartlett's work turned an ethnological question into a psychological one, and back again, in a manner very similar to the stakes of Rivers's work. Along the way, it offered the possibility of supplying a causal explanation, or rather a mechanics, for cultural diffusion.

Bartlett showed the text of "The War of the Ghosts" to a series of Cambridge students, staff, and townspeople about 1914 and 1915, with follow-up experiments over the next decade and beyond.[32] Each test subject was asked to read the story twice, at an ordinary reading rate. Bartlett did not assign a fixed time, because he said the significance of this period would vary from reader to reader. Then they attempted to write it out again word for word, after varying lengths of time. Bartlett examined differences between the original text and the reproduction, then tracked the sequence of changes as a single subject tried to reproduce the same story on several occasions. The earliest reproduction was done the next day (perhaps some fifteen hours later), the latest after ten years.

There were several kinds of narrative at work here. First there was the original story. Then there were the multiple narratives that subjects produced after reading the story and their reflections on the process of reading and reproducing it. Finally, Bartlett himself studied his subjects' report with a kind of narrative criticism, an approach that is intriguing given that he was a near contemporary of another Cambridge scholar who would be known for his technique of narrative analysis: I. A. Richards, who took the moral sciences tripos about the same time as Bartlett. Richards's criticism has been called psychology dressed as literary criticism; one might argue that Bartlett was repaying the disciplinary compliment.[33] Bartlett worked through the reproductions, analyzing changes and inferring the mental processes he thought had created them. He also drew on his subjects' reflections on the experiments—something that was most likely easier because several of them were his own students. For instance, when Bartlett claimed that people rationalized the story as they attempted to reproduce it, he noted that some of his own subjects thought so too. One person, reflecting on the role of the ghosts in the story, thought that "the main point was the reference to the ghosts who went off to fight the people farther on. I then had images, in visual form, of a river. . . . I saw the man telling his tale to the villagers. He was pleased and proud because the Ghosts belonged to a higher class than he did himself. He was jumping about all the time. . . . At first I thought there must be something supernatural about the story. Then I saw that the Ghosts must be a class, or a clan name."

Bartlett observed that this subject had missed the point about the ghosts, and he explained this by concluding that it was easier for this subject to remember the story if he could see the ghosts as a clan. Had he not "rationalized" the ghosts, he might have had to omit them.

In surveying the reproductions as a whole, he found that some details were remembered well (e.g., "something black came out of his mouth"). Others were changed to conform to subjects' expectations, as exotic elements gave way to features more familiar to the social group or culture the subject belonged to. And some elements were omitted entirely. Bartlett concluded that the mind develops culturally specific expectations about the parts of a story and the relation between them. Details from the original that conform to these expectations are retained, while others are "rationalized" or "conventionalized" to make them fit, and still others are omitted entirely.

In an analogy that Hugo Münsterberg would have found intriguing, Bartlett compared the mental work of the storyteller to the structure of an early film. In "The War of the Ghosts" there was a sequence of events, but the connection between them was not usually explained. This situation was similar to what "would confront the spectator of one of the earlier cinematographic films with the usual explanatory tags omitted." That is, it resembled a silent film with no intertitles to explain the relation between one scene and the next. His experimental subjects "supplied the tags but without realizing what they were doing." They added new connecting, contextual phrases that explained the relation between the pieces of information. The result was "denuded of all the elements that left the reader puzzled and uneasy, or it has been given specific associative links which, in the original form, were assumed as immediately understood." This raised the question of how to define a coherent narrative, and Bartlett had an empirical answer: narrative coherence was achieved—or perhaps approached—through the changes made over time to statements that were not initially meaningful. Since this was also what produced "conventionalization," it and narrative coherence were closely related.

Bartlett's experiments on memory would depict perception and remembering as tightly entwined. Remembered material was called on to make sense of new perceptions and could swamp a new perception; perception in turn could transform existing remembered material. But how could one explain the nature of these changes?

In the late 1910s, Cambridge was abuzz with psychoanalysis. One text in particular, Freud's *Psychopathology of Everyday Life*, provided a framework for interpreting verbal "mistakes" as meaningful constructions by the mind. Bartlett almost certainly read this work, which had been translated and published in London before the war. Bartlett's fellowship dissertation was full of terms common in Freud's early work—"projections," "consolidations," and "transferences." Such terms were common currency in Cambridge in the mid-1910s. It would be a mistake to treat Bartlett's work as being Freudian in the sense of a work like Rivers's *Instinct and the Unconscious*. But his general claim—that personal motivations, contexts, and preoccupations unconsciously change what we perceive, remember, and express—was in all likelihood indebted to Freud.

Clinical psychology also shaped Bartlett's approach, even though he never became a clinician. Bartlett later reflected that his early mentors' clinical orientation was important to how he understood his own work in the 1910s. Myers, Rivers, McDougall, and Head were all "doctors of medicine, interested in diagnosis," and all "preferred to treat human reactions as wholes rather than by detailed analysis."[34] And in 1916 Bartlett got a taste of this work too, when he was asked to help out with "shell shock" casualties.[35] He had no training in psychiatry; apparently "a psychologist" was all that was required. He later wrote that he thought patients might not have gotten much out of his efforts, but he himself was changed by the experience. He came to value collaboration with doctors, and interdisciplinarity more generally:

> There was no properly trained medical personnel available for dealing with these cases, and as a psychologist with practical interests I was invited to see what I could do to help them. I had discussed current treatments of "shell-shock" with Myers and Rivers and, like all psychologists of the period I had read everything I could get of the work of Sigmund Freud. . . . I cannot say whether what I was able to do was of much, or indeed of any, use to these

patients: none of them was allowed to stay long in Cambridge. But I myself learned two lessons. ... The first concerned the value to the psychologist of collaboration with medically trained experts, and the second that psychological insight and understanding always demands a consideration of the internal, and personal, conditions of behaviour as well as a study, and if a possible a control of external circumstance.[36]

Although his interest in Freud eventually soured and the program he later developed in Cambridge would stay aloof from clinical psychology and psychoanalysis, he continued to maintain that his experience with shell shock patients taught him not to try to isolate a single aspect or component of mental functioning. One needed to find a way of studying mental functions as part of a complete system.

Transformations

Bartlett wrote up his research in a 1916 fellowship dissertation that earned him a position at St. John's College. It chronicled his experiments on visual perception and argued for the powerful role of a "preformed" scheme or set of assumptions in structuring incoming information. The dissertation opened, however, with an evocative quotation from Robert Louis Stevenson: "'Faces have a trick of growing more and more spiritualised and abstract in the memory until nothing remains of them but a look, a haunting expression; just that secret quality in a face that is apt to slip out under the cunningest painter's touch.' He is giving an illustration of something which almost everyone must have noticed, but which hardly anybody has systematically studied."[37]

Stevenson was not merely saying that memories of faces faded over time, Bartlett explained: there was something distinctive, "that secret quality," in what was retained. Whatever this quality was, it was enhanced as other elements faded. It was also noteworthy that the passage was about a putatively ordinary experience. Bartlett was interested in routine mental attributes, not distortions created by the artifices of the laboratory.

What Bartlett chose *not* to quote is also worthy of note: he omitted the rest of Stevenson's reflection, that "we expose our mind to the landscape (as we would expose the prepared plate in the camera)." The analogy of a photograph belonged to a convention of memory that Bartlett had set himself against, the idea that what we observe makes its record in our minds as an impression stamped from without. Photography was not useful even as a foil, because, unlike Münsterberg, Bartlett did not want to measure memory against an ideal of faithful record taking. He wanted to adopt an entirely different set of references. These became clearer in a stream of papers he published in the 1920s, and eventually in his famous book.[38]

During these years, remembering was not Bartlett's ultimate goal. It was a conduit to an understanding of conventionalization, of how new cultural material is adapted and made meaningful in new contexts.[39] He saw his work as explaining and potentially redefining conventionalization, working in the laboratory rather than the field. He toiled away on his long-deferred book, though he struggled with the topic. At length he decided to make recollection the focus of the book's title, though his Cambridge colleagues probably would have preferrred conventionalization.

Remembering

The writings Bartlett is now most famous for did not appear until more than fifteen years after these experiments. In the meantime, the institutional and intellectual context of his work changed enormously. The book he published in 1932, *Remembering*, bears the marks of this history, becoming something like an archaeological site that retains layers from different stages of his career.

At the end of the war, Bartlett's senior colleagues returned, but not for long. (Myers resigned shortly after his return; Rivers died in 1922.) Bartlett became chair of a new department of experimental psychology just as Cambridge was entering a period of intense development, and he presided over the launching of the university's first graduate program in psychology.[40] During these years he published a stream of articles, initially drawing almost verbatim from his fellowship dissertation[41] but then branching out into a more intensive engagement with anthropology. Even as he looked outward, beyond the individual, to reflect on the interaction of groups, he also sought new tools for looking inward, to ask not only what kinds of changes mental processes made to incoming information, but why and how they did this.

One challenge Bartlett faced was to take a stance on psychoanalysis. Like many of his colleagues, he had been intrigued by Freud's writings in the 1910s. In 1920, Bartlett staked out a carefully qualified position in an argument about folk stories—a natural topic after "The War of the Ghosts." Several psychoanalytic writers had used the tools of dream analysis on folk tales and myths, arguing that themes common in folk tales were the expression of wishes and complexes fundamental to the human condition. For instance, Otto Rank had explained certain folk tales as emerging out of unconscious psychosexual life. Rank argued that the recurrent features of folk tales showed they were universal. Bartlett rejected the assumption that all minds are fundamentally alike, so that all symbols, dreams, and tales can be interpreted the same way because they "originate from the same spring, the human psyche."[42] He particularly disliked the notion of a "wish" as a self-evident category that could be identified without contextual information; Bartlett thought it required complex analysis and research. What was missing in psychoanalytic analysis generally, he claimed, was an attention to context and contingency. Two somewhat similar-sounding stories might have quite different meanings when one came to know more about who told them to whom and why; the problem was particularly acute when symbolic material was involved, as it often was in folk stories.[43] Sociological and anthropological approaches did attend to context, but they did not address the individual.[44] Bartlett's own work, he hoped, would strike a middle ground, forging a synthesis of the contextual, the social, and the psychological.

Much of Bartlett's work of the 1920s sought to deliver on this promise, particularly *Psychology and Primitive Culture* (1922).[45] It is now known for its rejection of the claim that "primitive" thinking is "pre-logical,"[46] but Bartlett also used it to make his psychological work on conventionalization relevant to anthropologists. He argued that as cultural forms traveled they were subject not only to simplification, but also to elaboration. Here the constructive character of meaning he had documented in the 1910s was applied to collective knowledge. More generally, the "diffusion of new culture" depended on the prevailing culture of its new community: "The elements introduced must be attached to a demand which is widely spread throughout the community concerned." The fate of a cultural convention was in the hands of future users.

In 1932 Bartlett developed this idea into a section of *Remembering* titled "Social Psychology." Although it began with an appreciative reference to an influential work on social psychology, it mainly cited anthropological work and was framed around the idea of collective memory.[47] Was there any sense in which a group could have a memory? This was not long after Carl Jung had articulated his idea of a "collective unconscious," and the question of how a group could know or remember was being passed around in psychological discussions. Bartlett thought it was not simply that what seemed like a common memory shared by a group was the aggregate of many individual memories, added up or averaged, as it were, because a group was not simply a collection of people. It had an organization, and this provided a context for any individual act. Memories that people articulated "for" their group were affected by their place in the group and were therefore different from memories of their own individual experiences. Bartlett then discussed how salient characteristics of groups could affect the recall of "group" memories, adducing Rivers's and Haddon's concept of conventionalization, but with an emphasis on the idea of "social constructiveness" that he had developed in his work on individuals (and also in other research on military technology during the war).[48]

As Bartlett struggled to find a way to engage with anthropology as a psychologist, he also tried to find a way to engage with other approaches to the functions of an individual's mind and brain. He found what he was looking for in the concept of "schema" developed by neurologist Henry Head. In the 1910s, Head had studied how the cortex handled information from the peripheral nerves, especially in tracking bodily movement. Before Head, the benchmark work was Hermann Munk's, in the 1880s.[49] Munk claimed that each movement left a memory trace in the brain, a picture of the position of the body at a particular moment. Each of these traces was static and self-contained. With each new movement, the trace corresponding to the previous movement would be reactivated. The brain kept track of movements by comparing each new picture with the previous one. Head had studied a series of patients who suffered from cortical lesions and found evidence to rebut Munk's idea of discrete "trace" records of bodily position. For instance, a patient "with a certain cortical lesion may be able to image accurately the position of his outstretched arm and hand on the counterpane of a bed." If he closed his eyes and his hand was raised (by someone else) and moved to a new position, the patient might "localise the spot touched on the skin surface perfectly well, but he refers it to the position in which the hand was, because he has entirely lost the capacity to relate serial movements."

Head was skeptical of the idea of a memory trace, at least in its conventional definition as a discrete kernel of information with no precoded connection to other mental content. He argued that neurological records of position came coded with comparative information. Munk had suggested that each trace was an absolute (as opposed to relative or relational) record of location and posture. The brain tracked movement by examining a pair of traces and comparing them (thereby, presumably, creating a third: the trace of the brain's comparative activity). Head maintained, instead, that information about position was *essentially* relational and contextual, with comparison built into each neurological record:

Every recognizable (postural) change enters into consciousness already charged with its relation to something that has gone before, just as on a taximeter the distance is presented to us already transformed into shillings and pence. So the final product of the tests for the appreciation of posture, or of passive movement, rises into consciousness as a measured postural change.... For this combined standard, against which all subsequent changes of posture are measured before they enter consciousness, we propose the word "schema."[50]

These schemata were not just disparate records, say, of the movements of a limb, but were complex representations of the whole animal. As we move, he argued, "we are always building up a postural model of ourselves which constantly changes."[51] Incoming sensations are not recorded in the brain in their raw state. Instead, schemata change them "in such a way that the final sensations of position or of locality rise into consciousness charged with a relation to something that has gone before." This amounted to a "model of the organized animal" rather than a massive quantity of discrete molecules of information. Head produced no clearer or more elaborate explanation of schemata than this, and he acknowledged that it was both vague and empirically underdetermined. The key point, however, was that information could not enter or exist in the brain except in a form that related to other information, and that "schemata" referred to these complex and plastic structures.

Bartlett was fascinated by the notion of the schema. Toward the end of an enthusiastic review of Head's book on aphasia, he made some general remarks relating this idea to the study of "thinking." Thinking, he said, possesses three general characteristics:

(a) It is a capacity for dealing with situations at a distance, and hence involves the use of symbols; (b) it is a capacity for responding to the qualitative and relational factors of a situation in their general aspect, and hence involves the formulation of symbols; (c) in the great majority of cases it is a capacity for retaining the concreteness of a given situation, and at the same time for utilizing these abstract and general qualitative and relational features.... We can watch this very complex capacity growing up.[52]

An organism's response to a stimulus depends, he wrote, on an "arrangement of preceding responses which have already been organized." These organized responses could have effects the organism was unaware of. The name for one of these organized responses was "schema." And at any level of physiological response, a number of these schemata must be in operation. Bartlett liked the idea that using mental images involved developing "a device for picking certain concrete items out of the settings in which they originally occurred, and for utilizing them irrespective of the precise sequential order to which they actually belonged in the first place."[53]

In *Remembering*, Bartlett acknowledged the commonsense appeal of the still commonplace "trace" theory of memory, but reminded readers that his 1914–16 research had shown that "the past operates as an organised mass rather than as a group of elements each of which retains its specific character."[54] One could begin to understand how this organization worked if one looked first to neurology, and specifically to Head's work.

Bartlett thus made "schema" central to his account of remembering. He jettisoned some parts of Head's definition, including the role of conscious awareness in the development of schemata. Bartlett believed that consciousness was unusual in most mental processes, but he retained the idea that all new information must be placed in the context of a well-organized existing framework in the mind. When material would not fit easily into that framework, either the framework would change or, if no accommodation could be made, material would be discarded. Such a model did not allow for any kind of simple "trace" understanding of memory, since there was no way sensations could "imprint" information on the brain without that information's being transformed. Rather, schemata helped Bartlett articulate his notion of an "active organized setting":

> Schema refers to an active organization of past reactions, or of past experiences, which must always be supposed to be operating in any well-adapted organic response. That is, whenever there is any order or regularity of behavior, a particular response is possible only because it is related to other similar responses which have been serially organized, yet which operate, not simply as individual members coming one after another, but as a unitary mass. Determination by Schemata is the most fundamental of all the ways in which we can be influenced by reactions and experiences which occurred some time in the past. All incoming impulses of a certain kind, or mode, go together to build up an active, organized setting: visual, auditory, various types of cutaneous impulses and the like. . . . There is no reason . . . to suppose that each set of incoming impulses, each new group of experiences persists as an isolated member of some passive patchwork. They have to be regarded as constituents of living, momentary settings belonging to the organism, or to whatever parts

of the organism are concerned in making a response of a given kind, and not as a number of individual events somehow strung together and stored within the organism.[55]

A theory in terms of schemata thus evoked the sense of context (an "organized setting") and also the idea of practice (the "setting" actively changed information coming into the mind). There was also a developmental, or perhaps even a social-evolutionary, dimension to the notion, because schemata were built up over time and adapted to challenges from new experiences. In general, schemata encompassed the idea that memories contain past experiences, but also that they are dynamic and constructed anew every time they are recollected. Bartlett used an example from tennis to illustrate how the past operates on new perceptions "as an organized mass":

> Suppose I am making a stroke in a quick game, such as tennis or cricket. . . . When I make the stroke I do not, as a matter of a fact, produce something absolutely new, and I never merely repeat something old. The stroke is literally manufactured out of the living visual and postural "schemata" of the moment and their interrelations. I may say, I may think that I reproduce exactly a series of text-book movement, but demonstrably I do not; just as, under some other circumstances, I may say and think that I reproduce exactly some isolated event which I want to remember, and again demonstrably I do not.[56]

The concept of schema, revised and applied to Bartlett's early research, displaced "conventionalization" as his central focus and replaced it with "remembering." Until well into the 1920s, Bartlett had thought of himself as studying conventionalization. The point of the research was different now: instead of asking how ideas become "conventionalized," his goal became "remembering." He was now focusing on the mind's actions, not on the resulting ideas.

Because schemata had a structure and an organization, new information that did not fit would have to be discarded or changed, or else the schema would have to change in response. In "The War of the Ghosts," Bartlett found that information that could not be made to fit was discarded. This suggested that schemata were applied to incoming information to make sense of it. But Bartlett also found that when this practice was inadequate to the task, individuals or groups could rework schemata to find new ways of using them, or could revise them to meet new purposes. The advantage of the notion of a schema over a "trace" model was that it could explain how habits could change. Schemata allowed an individual to revive memories that had been laid down at varying times, rework them, and form new combinations. Bartlett believed this was the purpose for which consciousness had evolved. The creative reworking of memories, their relation to each other, and their significance for future action were key functions of consciousness. Bartlett used a phrase for this that many readers found confusing: it was a process of "looking at" or "turning around on one's own schemata."

Remembering Remembered

Bartlett's memory experiments have been reproduced many times over several generations—a reflexive irony not lost on the many psychologists and historians who have written about them.[57] Bartlett's book itself has become almost iconic in recent years. According to some scholars, it is the single most widely cited psychological work of the twentieth century. Yet in the decades immediately after its publication, it did not have anything like this prominence. One of the main reasons for that neglect was that Bartlett's book appeared at a time when behaviorism was dominant, at least among American psychologists. Its attention to mental "content" and "ideas" rather than behavior ran directly against the grain. But just as important was that Bartlett's work was out of kilter with a rising trend in the representation of memory.

From about 1940, forensics experts, military psychiatrists and administrators, clinical psychologists, filmmakers, and others were increasingly portraying memories as stable, permanent entities that could be "revived" (if they were inaccessible) and in some cases even re-experienced given the requisite techniques. This became a foundation for the treatment of psychiatric casualties during World War II, as we have seen, and a widely accepted way of thinking about autobiographical memory in general. Bartlett was rediscovered in the late 1960s and 1970s only as that midcentury consensus disintegrated. Two important intellectual movements then seized on it and made it central to their approach to the construction of knowledge.

In psychology, a movement arose to replace the old behaviorist framework with a different approach according greater recognition to mental content, thinking, and meaning. Criticisms had been fermenting, nameless and without formal identity, since the 1950s. But in 1967 Ulric Neisser's massively influential *Cognitive Psychology* set the agenda for the new field. Neisser embraced Bartlett's memory work wholeheartedly, from the reproduction experiments to the concept of a schema. Most of all, he revived Bartlett's holistic and naturalistic approach to mind. His work made "schema" and "reconstruction" ubiquitous themes in subsequent psychology, as well as in associated fields such as the computational sciences.

Neisser and like-minded cognitive psychologists promoted the "practical" study of memory—what he later called the "low road" of naturalistic memory research, in contradistinction to the "high road" of traditional laboratory experimentation.[58] He wanted research to proceed in real-world settings, this being the only way to approach a subject that was intricately and essentially enmeshed in that world. The forensic psychology movement led by Elizabeth Loftus was one of these initiatives. Although only one of several such enterprises, because its legal significance soon made it the branch of psychology best known to people outside professional psychological circles. And for its advocates Bartlett's reconstructive account of memory offered an extremely useful explanation of how memories could seem authoritative yet not be trustworthy because, being continually reconstructed, they could not be in any simple sense "true." Bartlett thus came to be yoked to the "faulty memory" arguments that were the hallmark of forensic memory research.[59]

About the time that Neisser was installing Bartlett in the foundations of cognitive psychology, another quite different intellectual movement was building on a related aspect of his work. In the 1970s, a new field called the sociology of scientific knowledge emerged in England and Scotland. Its goal was to develop a sociological understanding of the development of scientific knowledge, based in a notion of "social constructivism." The impetus for this movement was similar in some respects to that for cognitive psychology. In insisting that theirs was a sociology of scientific *knowledge*, its principals were pointedly departing from an older sociological tradition that sought to explain the practices and institutions of science socially but assumed that the cognitive content of scientific work could not be explained in such terms. The new endeavor would extend to the knowledge itself—and not just to rejected or superseded knowledge, but to scientific ideas currently regarded as authoritative.

David Bloor, one of the framers of the field, called his version the "strong program," in contrast to the "weak" sociology that restricted social explanations to marginal science or false beliefs. The intellectual roots of this project were many, but among them was Bartlett's own notion of "social constructivism." Bloor had in fact taken a psychology degree in Cambridge in the 1960s, where he had been taught *Remembering* by Bartlett's students in the place where it was written. When he turned to the study of science, he wanted to study scientific thinking with some of the skills and perspective he gained in the psychological laboratory. As he wrote some years later, the basic idea behind the "strong program" was parallel to the interest of experimental psychologists in studying "the processes of individual cognition." In like manner, "we can generalize the enterprise and turn scientific curiosity onto collective cognition." Scientific thinking could not be studied merely through philosophical reflection or narrow analysis of one particular element. As Bartlett had taught about individual cognition, it had to be studied in context, in a "naturalistic setting."

Bloor set himself to do for scientific work what Neisser was just then urging psychologists to do in their study of individual thinking practices. He later wrote that he learned two "laws" from Bartlett (his terms, not Bartlett's), the law of "complexity" and the law of "conventionalization." The law of complexity taught him, as it did Neisser, to study thinking in a natural context. Bartlett's refusal to make a distinction between thinking processes that resulted in "true" and in "false" beliefs was also important to Bloor, who thought, like Bartlett, that the same cognitive processes underpinned intellectual claims that were highly respected and those that were marginal or rejected. Bloor later said that in the 1960s, when he read Bartlett for the first time, Bartlett's basic idea that "the complexity of a response depends on the complexity of the responding organism, not the complexity of the stimulus"—an idea that became central to the sociology of knowledge—struck him "as an incredibly powerful insight."[60] Bloor also found what he called Bartlett's second law—of "conventionalization"—useful as much for interpreting attacks on his program as for constructive reworkings (other scholars would have called them misunderstandings) of what he had intended to do.[61] Few people in science studies would have guessed at the importance of Bartlett for some of the basic principles underpinning the field—Bloor's colleagues knew nothing of Bartlett, and Bloor himself, although seeking to apply to science the approach to knowledge he learned in his psychology program, forgot that Bartlett himself had used the term "social constructiveness" (which became central to science studies) and wrote about it in a similar spirit.

By the 1980s Bartlett's work had taken on a curiously contradictory role in two of the most prominently contentious debates about the status of scientific knowledge then occurring, one in academia and the other in the legal system. In the sociology of scientific knowledge it had played a foundational role, but one that was largely unknown to the field's practitioners. In forensic psychology, on the other hand, its role was almost too well known. Memory experts liked to invoke Bartlett as providing a respectable genealogy for their skeptical attacks on remembered testimony in the courtroom, but the memory research they recalled was at best a partial representation of his original project.

IO

MAKING FALSE MEMORY

The surge in allegations of long-forgotten childhood sexual abuse during the 1980s sparked a period of intense controversy that came to be known as the memory wars. What made them so fierce was the emergence of a coherent movement skeptical about the memory claims. At first the opposition to "repressed memory syndrome" had been disorganized and lacked a theme. But from 1992 it became wedded to a directly competing rival concept: "*false memory syndrome.*" The new concept redefined the experiences of people who described an abusive past: their so-called memories were certainly powerful and consequential, but they were also false.

Repressed memory syndrome itself had been central to what might best be called a culture of traumatic memory recall—a culture because it was so complex, multifaceted, and demographically widespread. No single individual, group, community, or body of knowledge can be identified as a necessary or sufficient initiator of, or contributor to, the growth of interest in repressed memories of childhood sexual abuse. It developed in ways that were hard to track, being a subtle braid of psychiatric and psychotherapeutic research and clinical developments, trends in public culture, and grassroots therapeutic movements. By 1990, the recovery of repressed memories had become a national trend. As extraordinary as these claims were, it had become harder to disbelieve them. At that moment of apparent triumph, false memory syndrome appeared on the scene.

Unlike repressed memories, the beginnings of the concept of false memory syndrome are easy to identify. The term was coined in the early months of 1992 by Peter and Pamela Freyd, a couple affiliated with the University of Pennsylvania (Peter was a math professor; Pamela worked in the educational branch of the Center for Cognitive Science), in discussion with a small group of associates. One of the Freyds' daughters had accused Peter Freyd of childhood sexual abuse, the memory of which she had repressed and not recalled until late 1990, when she was in her thirties. The couple's response, after denying her claims and trying to persuade her that her memories were mistaken, was to form one of the most effective intellectual-social campaigns of the late twentieth century. They did not merely contradict the accusations; they took the argument into the national spotlight. Peter Freyd later explained that they realized that "if this is happening to us it must be a national problem, and it will require a national solution." The solution, they decided, would involve addressing the very concept of recovered memory syndrome. Peter Freyd later recounted that it was clear to them that it would not be enough to "point out that there were problems with the definition and diagnosis of 'recovered memory syndrome' as a psychological condition." They needed to be able to "point to a real condition, a counterdiagnosis with its own indicators." After much discussion of possible terms and definitions (they rejected "pseudo-memory syndrome" "because of the initials 'PMS'"), Freyd decided on "false memory syndrome." There were a few circumstantial indicators— "the involvement of a psychotherapist, the use of memory-refreshing practices like hypnosis, the sudden appearance of a memory shared by no one else."[1] Like repressed memory, false memory was linked to a syndrome, and it had its own loose cluster of indicators. But the indicators were very different, and the conclusions they to led were radically opposed.

The history of false memory syndrome is in certain respects even more interesting than that of the movement it arose to confront. It tells us not only about aspects of the history of memory, but also about ways that scientific legitimacy can be established, overthrown, and reconstituted. Issues that are routinely addressed in the rarefied circles of sociologists and philosophers of science were not only tackled in the courts but aired on *Oprah*, discussed in *Time* and *Newsweek*, and investigated on *60 Minutes*.

Reconstructing Recovered Memory

Abuse survivors frequently were advised that speaking out was the best route to personal recovery. A spike of litigation ensued, along with intense media coverage in magazines, trade books, and television. Repressed memory became the preeminent psychological topic of the late 1980s, and a defining feature of public debate.

But eventually there were protests that it all seemed too much. There was growing concern in the national press over the rise of repressed memory cases. The occasion for this closer and more sustained scrutiny was a series of spectacular trials in the very early 1990s, in particular that of George Franklin. Cognitive psychologist Elizabeth Loftus, who had made a career out of researching and speaking in court on the unreliability of witness memory, had hitherto ignored the new issue of *recovered* memory.[2] But she joined the fray in 1990 when she was asked to testify for the defense in the Franklin case. There she conceded that apparently repressed memories were powerful and felt real, but she nevertheless insisted that they were even more likely to be false than the eyewitness memories whose reliability she had impugned for years.[3] Other psychologists also spoke up, warning that the indicators for repressed memory, as advertised in self-help books, were far too lax, and that the numbers being touted for people suffering dissociative disorders (a defining feature of repressed memory) seemed far too high. After years of wide-eyed but respectful coverage of repressed memory, hints of skepticism also began threading through the news media. In October 1991 *Time* asked, "When can memories be trusted?" It noted that psychologists and lawyers "are finding that more and more cases turn on the question of how reliable memory is."[4] When, at the end of 1991 and early in 1992, the False Memory Syndrome Foundation launched itself into the battle, the groundwork had already been laid.

In December 1990, Pamela and Peter Freyd went to visit one of their daughters, a well-regarded research psychologist with tenure at a good university. During the visit she suddenly left the house with no explanation, taking the children with her. Her husband phoned her parents to tell them she had suddenly remembered being seriously abused as a child. They were asked to leave, and from that point the daughter cut off almost all communication. The family remains estranged.

The Freyds say that their initial response was panicked soul-searching. Could Peter Freyd have done these things and both Freyds forgotten? There had been problems during their children's childhood, including Peter's drinking, but they quickly ruled out the possibility that he had sexually abused either of their daughters. They began looking for information that could help them understand how such a memory could be false—and if it was, what they could do about it. Soon they resolved on an ambitious strategy. Instead of focusing solely on their own predicament—trying to persuade their daughter that she was wrong or seeking to reconcile with her without resolving the issue—the parents decided to focus their energies on the recovered memory movement as a whole.

Throughout 1991 the Freyds tried to research the idea of a coherent entity of false memories about childhood trauma. They had some prior interest and experience in other psychological issues—Peter Freyd regularly attended psychology colloquiums at the University of Pennsylvania, for instance, and at the time of the accusations, he was helping to found a cognitive science institute there; Pamela Freyd had earned a PhD on language acquisition in 1981 and had long been a member of the American Psychological Association. The connections both Freyds had with various research psychologists were to prove crucial in the months after the accusations. They seized on expertise wherever they found it, and they bonded with like-minded people they encountered along the way.

One of the very few books on false accusations of child sexual abuse that they could find in 1991 was written by Ralph Underwager, a therapist and seasoned legal consultant known for his skepticism about child abuse allegations. He and another therapist, Hollida Wakefield, ran an "Institute of Psychological Therapies" in Minnesota. They frequently consulted in custody cases involving issues of child abuse, arguing that false abuse allegations were often used to restrict parental access after a divorce. Wakefield and Underwager were disposed to be suspicious of claims of long-repressed memories of child abuse. Too often, they argued, such claims emerged from an embittered family situation, when an irresponsible therapist or lawyer became involved. Underwager gave the Freyds his support and became an ally for several years.

Through Harold Lief, a psychiatrist they knew from the university, the Freyds were then referred to social psychologist and clinician Martin Orne. Orne agreed to meet, but he insisted that Peter Freyd submit to a polygraph test (according to Freyd, it laid Orne's worries to rest). But Orne then became very interested in the issue of unreliable recovered memories, which he saw as an instance of "demand characteristics"—the same effect that had made him worry about forensic hypnosis, but this time operating on a massive demographic scale.

He thought recovered memory claims were the result of "confabulation,"[5] a psychological term dating back to the early years of psychoanalysis, referring to a process by which memories of past experiences were knitted together with fantasies or suggestions to create an imaginative construction that appeared to be a memory. Orne became one of the Freyds' most powerful and steadfast supporters. He helped them assemble an impressive advisory board, including some eminent scientists, the most prestigious being researchers in cognitive and social psychology.[6]

The Freyds also set about finding other families with experiences like theirs. They eventually assembled a small group (the Freyds, another accused couple, Orne, and Paul McHugh, then chair of psychiatry at Johns Hopkins), which met in November 1991 to form a strategy. Then they placed an ad in several magazines: "Has your grown child falsely accused you as a consequence of repressed 'memories'? You are not alone. Please help us document the scope of this problem." It gave a toll-free number that connected to Underwager's institute.[7] The Freyds and another couple financed the staff who answered the phone and prepared packets of materials that Underwager mailed to individuals who called the help line.

In a newsletter circulated in early December, the Freyds further speculated on how to develop the most effective network of individuals and groups:

> Perhaps we could operate at two levels—a national level in which we make every effort to document the scope of the problem by collecting research information which will remain totally confidential and a local level in which groups form as people act to meet each other. We (Institute and I) could expedite the local connection of people if we have a "contact" people in various areas.[8]

They brainstormed about what kind of organization was needed. Peter Freyd later said he wanted to emulate the American Heart Association—to promote research on recovered memory and popular appreciation of its hazards. Financial support came in from a wealthy couple who had also been accused of abuse, and the Freyds set about establishing a nonprofit body. A few months later there was a national organization with a modest salary for Pamela Freyd, a budget for other costs of developing and running the organization, and a rising number of regional and local groups.

In January 1992, amid a discussion of how to define the nascent movement, someone proposed that "Forum on Real and Imagined Sexual Abuse" would reflect the purposes of the various people drawn together by their shared concerns about recovered memory. But Peter Freyd argued that there needed to be a specific entity—a disease, a syndrome—that could be the central concern of the research foundation. Researchers and journalists would need such an entity to focus on. The entity "false memory syndrome" therefore came into formal existence alongside the institution that sought to eliminate it: the False Memory Syndrome Foundation (FMSF). Like recovered memory, it had its own researchers, institution, and moral charge. It also had its own checklist of symptoms: the involvement of a psychotherapist; the use of practices like hypnosis, guided imagery, or word-association tests; the sudden appearance of a memory that no one else in the family (and no medical records) could corroborate.

The false memory movement grew in a way that looked organic and unorchestrated from the outside. But the Freyds did not think of their nascent operation as a grassroots organization, however much it seemed to arise "from the ground up." Peter Freyd said in retrospect that they considered themselves a "scientific establishment in waiting." That is, they meant to create an intellectually elite set of beliefs. This new orthodoxy was waiting to emerge from its chrysalis into its appropriate role, aligned with scientific authority. The "voice" of the FMSF was therefore an unusual mixture of the authoritative and the grassroots. It aspired to orthodoxy, making impersonal statements appealing to the power of scientific research, but it could also take the emotional, sometimes strident, tone of grassroots organizations. It also published personal testimonials and articulated goals of mutual support and public consciousness-raising.

In 1992 and early 1993, local meetings of skeptics in this emerging movement were advertised for Maine, New Hampshire, Seattle, Missouri, New York, Ohio, Alaska, Utah, and California. There were also regional "tristate" meetings. The names and the origins of many of these smaller organizations suggested an association to the FMSF, but the Freyds did not want to encourage perceptions of an organized network run by the foundation. They saw the FMSF as a research entity, documenting a mysterious pattern. In contrast, local and regional organizations were straightforwardly focused on mutual support. The motto of the Texas FMS organization, for example, was "hearing-hoping/helping-healing"; the Virginia group's motto was "families and professionals networking together."[9] Every month the FMSF published a geographical list of families who had contacted them. A list from the spring of 1992 included 243 families in twenty-five states, and three in other countries. By late November the total had risen to 2,010, and by the end of 1993 there were several thousand families on the list.

Supporters saw the Freyds' ambitious formulation of false memory as heroic, even self-sacrificial. The attendant publicity could surely be expected to make reconciliation with their daughter harder, at least in the short term. And they would certainly find themselves the targets of opprobrium, suspicion, and worse. Advocates of recovered memory therapies and survivor groups saw them as self-serving. They charged that it was in the interests of culpable parents to locate the problem in the therapeutic culture rather than in their own actions. This, they alleged, was the true goal of the FMSF.

The Campaign

Once formed, the False Memory Syndrome Foundation got to work with the intensity of a national political campaign. Pamela Freyd recounted how they talked "to anyone who would listen." But even this is an understatement: FMSF members talked, wrote, pleaded, and shouted not only to anyone who would listen, but also to anyone who wouldn't—and they threatened to sue at least one or two of those. Pamela Freyd assembled information packets that went out to everyone who contacted the FMSF. These packets were sent to families, to therapists, to politicians, to scientists, and to every branch of the media. Each packet supplied everything a reporter would need to assemble a story friendly to false memory. The packets also included samples of material supporting recovered memory but, as one would expect, there were fewer materials on this side. (The Freyds had been advised by a local reporter on how to assemble a standard press packet, but the model Pamela Freyd adopted was, according to Peter Freyd, more like "the package of supplementary material distributed at the beginning of a graduate course.")[10]

As these packets flowed out, news clippings, videotape, and other materials streamed in from across the country. These were carefully logged into a database, evaluated for usefulness and errors, and put to work. They were sent off to scientific advisers and others who might be able to use them as a springboard for research, as ammunition in some local controversy, or to document the status of the issue in some particular locale. In many cases, particularly if an article took a strong stand either for or against the false memory propositions, the FMSF would mobilize its members and associated scientists to write to the journal with praise, censure, or corrections of mistakes, however slight.

Contemporary commentators on the memory wars sometimes spoke of recovered memory as a cancer afflicting American society. If one were to take up this metaphor, the FMSF's response was less keyhole surgery than full-body radiation. The FMSF was not merely a support group, not merely an advocate, not merely a popularizer of scientific knowledge or an advocate for research. It was all these things at once, on steroids. It is hard to think of precedents or parallels except perhaps the movement surrounding AIDS.[11]

Although figures within the FMSF did not like to portray it as a grassroots group, it in fact became a model for all such organizations of its kind. When accused families began to contact the FMSF, it kept records of names, phone numbers, and addresses. FMSF workers sent each caller the names of contacts within a few hundred miles, with information on how to reach them, along with sheets of guidance about how to start a local organization. Soon there were several regional FMS organizations, each holding meetings that were advertised nationally through the national FMSF. From these, in turn, came local groups. In a few months, more than a dozen regional groups had been founded, and membership had reached several hundred; at its peak, the membership of the FMSF stood in the high four figures. Its list of "affected" families numbered in the low ten thousands.

These member families became an active corps, heavily represented in the Northeast and on the West Coast but with members in most states. The FMSF gave them clear instructions. They should send the "the names and full addresses of your local newspapers, radio talk shows, and appropriate local television shows," from which the FMSF would organize its media campaign. They should add the names of "psychologists, social workers, lawyers and writers" to the mailing list, publish the FMSF toll-free number in community newsletters, and (after asking permission) put reading materials in churches and public libraries. At the same time, they were to forward articles about the memory debates to the FMSF so that it could "respond as an organization."[12]

The Freyds themselves, and particularly Pamela, did not work solely at their Philadelphia headquarters. In 1992 and 1993 she spoke at dozens of events. Some of them were very supportive—local meetings arranged by families who met each other through the FMSF. But Freyd also sent in proposals to psychotherapy and social work meetings where the legitimacy of recovered memory was taken for granted. These were liable to be anything but friendly: participants usually assumed that Freyd had served as an "enabler" of her daughter's abuse. She also spoke to meetings of journalists, senior citizen communities, judges, and litigators' groups. In short, she went anywhere and everywhere she could think of. She was not dissuaded when she took the podium only to see an entire roomful of people get up and leave. Just paging through the fifteen-year-old records of her movements in 1992–94 is exhausting.

Throughout the mid-1990s Pamela Freyd continued to be the engine of the FMSF, an indefatigable manager and traveler. She persevered even in the face of negative evaluations by audiences who had no ideological ax to grind. For instance, audience members at one continuing education seminar were enthusiastic about the topic but found that her talk did not meet their expectations; they thought she was too much of a polemical advocate rather than a "scientific" speaker. In short, she did not let her lack of expertise get in her way. FMSF members saw this as heroic, while abuse survivors took it as yet more evidence that Freyd was a partisan leader of a partisan lobbying group, not a spokesperson for a scientific organization.

Alongside its public efforts, the FMSF also carried on a more private proselytizing. Its core activity was one-on-one discussions with people struggling with memory problems—theirs or a family member's. In 1992 it fielded hundreds of calls each month from families who told "the now all too familiar story of a child (mostly daughters in their 30s) who suddenly recovers memories of abuse during therapy, of a child who confronts (often with a letter) and then refuses all contact with the parents."[13] Volunteers' most important task was to talk with members of "affected families," first on the phone and then sometimes in person. Most calls came from accused parents, but some came from people who had remembered abuse. Pamela Freyd described one dispiriting conversation with a woman who had telephoned the foundation doubting her own memories of abuse. The FMSF sent her some material, which she discussed with her therapist. When Freyd later called to check on her,

> She told us that she was certain now that her own therapist had not led her in any way and that the therapist had used hypnosis only once at the caller's request. Indeed, this person told us that she was becoming more and more convinced that her memories of satanic ritual abuse were real memories. She told us that since our first chat, she had validated her memories. She said that she had read stories in the newspaper and had seen things on television that proved to her that her memories were real.

Freyd added to this report her own warning that such reports of abuse could be "contaminated" by media coverage of recovered memory syndrome.[14]

A National Debate

The spring and summer of 1992 saw intensifying debate over the status of adult memories of childhood abuse, sparked by the FMSF. It organized a series of conferences on memory, giving a prominent platform for researchers skeptical of repressed memory. Media producers were invited to these conferences and given press packets referencing particular publications, events, and talking points. Similar packets went off to the deans and department heads of hundreds of social work and psychology programs.[15] One CBS network producer attending a meeting in Toronto reportedly exclaimed, "I had no idea that this phenomenon was so extensive!"[16] That November and December, major TV networks ran several documentary sequences on the FMSF. These were the beginning of a blitz of television news coverage.[17]

During this time the foundation grew rapidly, not only in membership (which doubled over the summer) but in its public visibility. A year earlier it had been awkward even to suggest that an accusation of sexual abuse might be wrong. Now articles in national newspapers were discussing just this possibility.[18] "Things certainly are heating up!" a visitor to the FMSF reportedly commented on the hectic activity in their office.[19] Freyd congratulated members that their efforts were paying off. The FMSF had formed "a team spread across the continent," with more than thirty volunteers in Philadelphia and dozens of workers nationwide writing letters, appearing on television, and giving interviews. Freyd later described it in the following terms: "At the peek [sic] of the time that publicity on talk shows was generating thousands of calls in 24-hours, we have [sic] about 125 volunteers around the country to help in responding to those calls. We had an assembly line going in the office to get out packets as soon as we heard back from the volunteers or for those calls we took ourselves."

These families became an active corps, voracious for actions they could take, particularly any that could be done while preserving their privacy.

The FMSF and its associated groups not only operated as a public campaigner but also provided members with advice, support, and an outlet for expression. They became at once political activist centers and parent support groups. From Philadelphia the FMSF printed letters and answered members' questions. It alerted them to theories of memory, including "confabulation," referring them to scientific literature on the subject and to writings that cast doubt on key concepts underpinning the traumatic memory theories related to recovered memory—concepts like repression, multiple personality disorder, posttraumatic stress disorder, and even relatively innocuous notions like flashbulb memories. It continued to provide contacts and advice to members who were being sued by their children,[20] and it gave updates on relevant legal cases and trends.[21] Indeed, the FMSF gave parents instructions not only for responding to their children's lawsuits, but for initiating their own. It provided information about third-party suits, in which parents sued their children's therapists for damages. The motivations were eerily similar to those of their children: litigation would be a form of therapy for the whole family. Only a full confronting of the therapist's quackery would show their children the truth. It was therapy for society, since suing one therapist would make others more cautious in the claims they made.

So in some ways the "false memory" community was not so dissimilar to the "recovered memory" networks it opposed. Both saw themselves as survivors of a terrible private injustice, one that was misunderstood by the wider society and even by many in the scientific community. Both saw themselves as possessing knowledge about memory (both their personal memories and a knowledge about how memory worked) that they needed to communicate to the wider society. Although these groups subscribed to opposing theories of memory, they nevertheless used some of the same language, although with striking differences of meaning. For instance, one parent lamented a daughter's "dissociated" state. For the daughter, who was presumably well versed in the literature on posttraumatic stress disorder and traumatic repression, this term would have referred to an unusual state of mind. But the mother used it to refer to a social state, a social and emotional estrangement from her closest community, "from many of her friends, her relatives, her husband's relatives and my husband and me."[22] The FMS community likewise told stories of recovery in relation to memory, but these now related to adult children who rejected their "false" memories and recovered from a period of delusion. While the survivor communities were outraged to hear about the existence of the FMSF, an organization they saw as inspired by an antagonism to them, the FMSF came to see the situation in a similar fashion, reporting in early 1993 that a Utah survivor support group called itself Survivors of FMSF. And while the FMSF regretted that the "current social climate is one in which 'therapy' has become political action" and that self-help books like *The Courage to Heal* recommended that readers "get smart by suing,"[23] it also urged its own members to get therapy and if necessary to take legal action (in some cases bringing suit against therapists) as part of their recovery from the damage inflicted by recovered memory therapists.

The Presumption of Innocence

One of the most obvious, pressing, and—one might think—troubling issues surrounding the False Memory Syndrome Foundation was whether this organization, and the concerns it explored and promoted, provided protection for people who were indeed guilty of child abuse. In early 1992, Pamela Freyd addressed this issue in a newsletter. At the end of a long discussion about why she was not worried that her group might serve as cover for abusers, she concluded,

> I believe in the assumption of innocence. I know that the stories I am hearing are silly. (Too bad that the people who wrote the script [for these news stories] didn't do more research.) I know that most of our children's therapists have assigned or recommended *Courage to Heal* and that this is one of the most biased books I have ever encountered in my life. I recognize that I am meeting some of the most delightful people I have ever met in my life, and I trust my judgment. I do not have the burden of proof. I hope no one ever again asks me how I can be sure that the families "didn't do it!"[24]

One year later, she wrote a less dismissive reflection on why the FMSF did not "screen" its members. She reminded readers that the FMSF had long disavowed the possibility of determining the truth of the stories it heard. As a research foundation, all it could do was "simply record and look for patterns." But that did not mean the FMSF counted and reported every story it heard. It tried to restrict itself to stories that involved "recovered" memories and where the existence of abuse rested wholly on the contested memory. The foundation did not, for instance, include cases in which there was acknowledged abuse but a particular event was in question, or those of parents whose underage children had been removed by the Department of Child and Family Services. The FMSF, she wrote, had too much to lose to "harbor perpetrators": "Are there mistakes in our list? We are human beings working in a situation of uncertainty and ambiguity. Of course there must be some errors. When resources permit we will be able to go back to records and determine error rates—if it is ever possible to know such things."

In early 1993, Freyd reflected again on the first year of the foundation. She listed the accusations of its critics that the FMSF was "a PR front for perpetrators," that it was part of "a satanic conspiracy," and that it was supported by "insurance companies." She noted that "we are a public charity and our books are a matter of record. . . . No insurance company has offered us any money." On the charge that the FMSF worked against feminism, she reminded readers that "approximately one third of the accusations of active abuse are against mothers." The insinuation that the FMSF was "part of the pornography industry" she seems not to have thought worthy of rebuttal. When some survivor communities promised to "storm" its "walls," Freyd sent a polite, mildly phrased note reassuring them that there would be no need: they were "cordially invited" to visit the FMSF's "modest offices" whenever they liked.

Trauma survivors and memory therapists were clearly alarmed by the FMSF. Some fears were extreme: that the foundation was a conspiracy of perpetrators lobbying to protect organized networks of abusers. Others were more modest: that the FMSF would shift the evidentiary grounds for investigating claims of abuse so that survivors, for whom it was already hard to come forward, would find it even more difficult in the future. Freyd denied any connection between the FMSF and the investigation of current instances of child abuse. "We have nothing to say about children or believing children," she insisted. The FMSF was solely about "adults and the techniques some therapists are using to 'help' adults find memories." However, the foundation did accept some connection between adult testimony and similar issues surrounding children, at least in terms of the kinds of literature that was worth collecting in the FMSF library. It archived a wide variety of documentation on child abuse testimony as well as adult survivor claims—and it amassed a large sheaf of literature on child abuse accusations that seemed to have been constructed, almost from whole cloth, as a result of aggressive and suggestive police and psychological interviews.[25]

Despite their hard work in distinguishing the aims of the FMSF from the issue of current child abuse, the Freyds were less successful in erasing doubts about their agenda than they were in creating doubts about "repressed memories." One reason was that their early ally (and FMSF founding member) Ralph Underwager, whose Institute for Psychological Therapies had provided the first telephone services for the nascent FMSF, was revealed to have given an extensive interview to a pedophile magazine, making statements that were broadly interpreted as supporting pedophilia.

In the interview with *Paidika, the Journal of Paedophilia,* Underwager was asked if pedophilia could be a responsible choice for individuals. He replied, "Certainly it is responsible." A series of questions and answers built on this theme, culminating with Underwager's assuring the *Paidika* interviewer that pedophiles could, "with boldness," proclaim, "I believe this is in fact part of God's will." He also compared the spiritual mission of Jesus with pedophiles' efforts to legitimize sex between adults and children. When asked if decriminalizing pedophilia was a legitimate goal, he replied, "Oh, yes." Pedophiles needed to "take the risk, the consequences of the risk, and make the claim: this is something good." He continued: "Paedophiles need to become more positive and make the claim that paedophilia is an acceptable expression of God's will for love and unity among human beings." However, he also cautioned, "whether or not they can persuade other people they are right is another matter." He warned that no American mainstream magazine would publish an argument that pedophilia could ever be healthy. Together, Underwager's statements combined an apparent moral relativism, a "constructivism" about sexual identity and, as he put it, sociopolitical realism.

After the *Paidika* interview, Underwager was publicly accused of supporting pedophilia. He protested that "radical feminists" claiming to be "sex-abuse experts" had taken the interview out of context: he had not been proclaiming pedophilia as part of God's plan, he claimed, but merely acknowledging that pedophiles could make such proclamations themselves. He was not defending their actions, only their freedom of speech.[26] But he resigned from the advisory board of the FMSF in 1993.[27]

Critics of the FMSF seized on the Underwager interview as evidence that the FMSF embraced child abusers, but it was probably more a sign of the foundation's initial haste and haphazardness. In the first year of their work, the Freyds were at the center of an intellectual, social, and political project that grew explosively. But perhaps because of the speed of this growth, they were not as careful to scrutinize prospective allies as they would later become. Their early efforts involved a kind of mad scramble for information, sometimes with the bar set low for forging alliances.

The Freyds had seized on Underwager and Wakefield as a lifeline—they had written one of the few books the Freyds could find about false allegations of abuse, and it said something they wanted to hear. Had they looked closer, they would have found that Underwager was becoming well known in 1990–91 for highly partisan expert witnessing for the defense in child abuse cases. He would argue that the testimony of children as old as ten or eleven was undependable, both because their memories were less stable than adults' and because they did not yet understand what it was to tell the truth. In 1990 he had been involved in a highly controversial case in Australia surrounding a notorious children's television personality, "Mr. Bubbles," in which Underwager was accused of misrepresenting research publications and, in essence, of making things up.

This suggests that the founders of the FMSF were forming neither a support group for pedophiles nor—despite their claims to the contrary—a scientific charity. They were establishing an interest group: interested, that is, in using any resources they could find to make the case that repressed memories were not real. Underwager turned out to look less like a respectable researcher than a disreputable hired gun who spent most of his time defending people accused of abuse or providing them with therapy. Other supporters turned out to be highly prestigious academics who not only had a respectable track record of research in areas relevant to the subject but developed new research as a result of their acquaintance with the FMSF. All these people claimed that repressed memory claims were not to be trusted. However, they developed their claims from different beginnings and intellectual contexts, used different evidence to support them, and related them to different contexts and individuals. The FMSF worked feverishly with what allies it could find to get its agenda a public airing. Yet it also aspired to be a disinterested scientific and philanthropic entity. It was the combination of these two identities and goals that provided the key to the FMSF's success. But it never really reconciled them, and that was a major reason the organization did not ultimately attain the kind of legitimacy to which it aspired.

Revenge of the Repressed

In 1990 a California businessman named Gary Ramona was sued by his daughter. She asked for damages in a civil case alleging long-past incest remembered only recently through the intervention of a therapist. She ultimately lost. But before the case was resolved, Ramona decided to bring his own lawsuit, not against her, but against the therapist. Other attempts soon followed. These new "third-party suits" became the skeptics' answer to the recently established legal culture supporting the concept of recovered memory.

Holly Ramona had been treated by psychotherapist Marche Isabella in 1989. Isabella gave Ramona the old truth drug sodium amytal in order to recover memories of suspected abuse. After the memories returned, Holly Ramona decided to sue her father. The initial suit was unsuccessful, but this was not the end of the matter. The accusations cost Ramona his marriage, his job, and his reputation.[28] He blamed her therapist. In 1991 he brought suit against Isabella and her medical center, alleging that their irresponsible treatment of his daughter had injured him. This was a third-party suit—that is, an action brought not by the person directly affected. Such actions were almost never successful. Ramona had to show that the therapist had a responsibility to him, and he also had to gather evidence that was usually confidential and almost impossible to obtain. But in this case the evidence *was* accessible. To bring her own suit, Holly Ramona had had to waive the confidentiality of her medical records.

The jury believed Gary Ramona, making legal history in the first case of its kind. Before dismissing the panel, Judge W. Scott Snowden called the verdict "extraordinary." Speaking to reporters afterward, the jury foreman, Tom Dudum, said he had been troubled that "she came in looking for treatment for bulimia and depression and ended up talking about something else. I say, let's work out the bulimia and depression and then move on to other things." The crucial part of this statement was the term "something else." Ramona's therapist claimed that the bulimia and depression were not "something else" from incest—they were in fact the body's code for it, and the whole point of therapy was to learn to read the code. So the jury did not just fail to follow the inferential path taken by Holly Ramona and her therapist, they rejected the basic analytical framework for making the inference in the first place.[29]

The Ramona decision was a new twist to the legal status of recovered memory. In a move that recalled the Washington court's rejection of therapy's "narrative truths," the court held Marche Isabella to the standards of the courtroom, since the knowledge generated in the consulting room could be taken out and applied, among other places, in a court of law. Here the court's machinery for evaluating the past was applied to therapist's techniques for evaluating the past—and found them wanting.

Commenting on the Ramona verdict, Zachary Bravos, a lawyer from Wheaton, Illinois, called it "a fundamental extension of tort law." Bravos had an interest in the result: he represented a woman who was suing her former psychiatrist, who had diagnosed her with three hundred personalities and, Bravos claimed, encouraged her to remember satanic ritual abuse. On the other side, Mary Riemersma, an attorney who represented the California Association of Marriage and Family Therapists, warned that the verdict opened "a Pandora's box," forcing therapists into the new and unwelcome position of "investigator trying to separate fact from fantasy." If critics like Riemersma were to be believed, third-party suits were forcing a change on the interpretative framework of therapy. Isabella herself warned that the verdict would make incest victims feel they had no one they could confide in.[30]

The Ramona case made therapists responsible not only to their clients but to the emotional circle around the clients. As Bravos put it, "These therapists might have to consider not only the patient, but other people central to the patient's life." Marche Isabella's lawyer, Sharon S. Chandler, predicted that this was why the verdict could have a chilling effect on therapists: a single patient represented the possibility of "6 to 20 potential lawsuits," from family and friends whose lives could be affected by the memories that emerged over the course of the therapy. Chandler warned that "anybody can barge in and say, 'If you hadn't interfered, my love relationship would still be going on or my child would still be obedient.'"[31] The ruling involved, therefore, a powerful assertion about the relation between individual and collective memory.

Psychotherapists had hitherto assumed that their responsibilities lay solely with their individual clients—that it was possible to explore their patients' lives without having to take into account the significance of their personal memories to anyone else. Of course, they understood that the self-knowledge developed in the consulting room would change their relationships with others, particularly their families. But it was assumed that psychotherapists did not have formal obligations to the wider world surrounding their clients. The Ramona decision asserted that they did. It was now impossible to separate the individual autobiography being developed in the consulting room from the collective memory of the family, or indeed the broader collective memory

Doonesbury comic strip, mocking recovered memory therapy. *Doonesbury* © 1994 G. B. Trudeau. Used by permission of Universal Uclick. All rights reserved.

Dilbert comic strip, 1994, mocking repressed memory. Reprinted with permission.

of the patient's wider community. The ruling legitimized a popular assumption that all these kinds of memory must align before their representation of the past could be considered "true." And if they did not, the therapist could be held responsible for injuries.

Ramona's was the first case in which an accused parent successfully sued his child's therapist; more such cases would follow.[32] Beyond litigation per se lay an increasing sense that there was a disturbing lack of accountability in what one might call the culture of therapy. Some therapists encouraged an understanding of selfhood and personal development that leaped to the inference that a struggling adult had been damaged long ago and that the person responsible could and should be identified and made to pay.

Cartoons alluded to the rise of an opportunistic trend of "victimhood," in which people portrayed themselves as significantly damaged by small slights from others and common, apparently innocuous personal experiences were read as signs of long-ago injuries of some extreme kind. There were also claims—some explicit and strident, some indirect, that repressed memory was a form of both individual and mass hysteria: individual memories were constructed through suggestion and emotional vulnerability, to the point that survivors had bodily symptoms that were psychological constructions. The idea was also contagious, as people looked at their neighbors and decided to join the traumatic memory bandwagon.

Another major blow to recovered memory came hard on the heels of the Ramona decision. In 1995, George Franklin's conviction was overturned by a federal appeals court. It concluded, among other things, that the trial judge should have given Franklin the chance to rebut his daughter's claims with newspaper articles on the murder. The possible line of defense was that Eileen Lipsker was remembering these articles, not events she had experienced herself. The court also affirmed that the broader issue of repressed memory

*"Now, just a darn minute, Fowler! You're not the only member
of this board who was a victim of ritual satanic abuse!"*

Comic printed in the *New Yorker* in 1993, lampooning copycat cases of traumatic memory.
© Lee Lorenz/The New Yorker Collection/www.cartoonbank.com.

was unsettled. The claims of repressed memory experts should therefore be
treated with ongoing suspicion, they concluded.[33]

The Ramona and Franklin rulings were not determining events in the
fortunes of recovered memory, but markers of how problematic such claims
had become. By 1993, however, the FMSF was confident. "It's happened,"
the foundation crowed to its members. "We can say it with confidence. The
climate surrounding the FMS phenomenon has shifted."[34]

False Memory and the Academy

If the False Memory Syndrome Foundation had been only a populist organization, it would probably have had little impact on the plausibility of
recovered memory. But it had an advisory board of outspoken scientists and
other academics who published research, issued press statements, and attended bespoke conferences on false memory. The FMSF's adoption of this
institutional form—the scientific advisory board—emulated the most suc-

cessful new scientific organizations of the period, namely biotech start-ups like Cetus and Genentech. Such corporations adopted big-name "scientific advisory boards" to lend credibility to their sometimes highly speculative scientific claims and (therefore) highly optimistic business plans. Success for these start-ups was measured in high share prices, profitable buyouts, and occasionally windfalls from initial public offerings. For a movement like the FMSF, the measure was the amassing of not financial but cultural credit.

But the strategy was more complicated than that analogy may suggest. The title "advisory board" implies that the scientists were there as a pipeline from the scientific community to the general public—that the role of the FMSF was to channel scientific knowledge from scientists outward into the culture, rather than to get members of the culture themselves involved in research. Unlike a biotech start-up, however, the parents as well as the scientists played an important role in defining this new syndrome. The most obvious examples were the FMSF's surveys and other practices of information gathering about their own members. More subtly, the members helped define an area of debate, shaping research questions and foci.

The first scientific conference the FMSF sponsored was held in April 1992, under the title "Memory and Reality: Emerging Crisis." This was neither an ordinary academic conference nor a meeting of "affected" individuals. It featured speakers from an extraordinary range of fields and specialties within academic and clinical psychology, psychiatry, and sociology, but it also included representatives from social work, the law, and the "affected families."[35] At the conference, the academics and professionals mixed with the families and "recanters"; all three groups gave papers. The atmosphere conveyed a sense of urgency and, perhaps inevitably, was far more emotional than was usual at scientific meetings.

The status of false memory syndrome as an underdefined but urgent problem made it "real" enough to command the attention of a wide spectrum of respectable individuals. Richard Ofshe and Margaret Singer, psychologists at the University of California at Berkeley, saw false memory syndrome as a standard "cult" attribute; it reminded them of Jonestown and the Hari Krishnas—though it should be noted that Ofshe seems to have found it fairly easy to define cultural and religious movements as cults, since in an affidavit only the previous year he had defined Sikhism this way.[36]

Martin Orne and his social psychology colleagues understood recovered memory as a powerful example of demand characteristics. The designation of false memory as a "syndrome" made it accessible to these multiple explanations, since the category conveyed reality and specified a list of characteristic features but did not specify a cause. It was plastic enough to be made to belong to any of several areas of expertise and, within these, a variety of research specialties.

Academics who attended this conference reported that meeting the parents intensified the urgency they felt about the repressed/false memory issue. And parents felt that the truth of their claims was being affirmed. The downside was that the scientists risked looking more ideological, because they were making common cause with parents (and not their children). Some saw their impassioned discussions as evidence of zealotry. But the alliance of families, academics, and professionals was emblematic of a growing trend in handling research areas of social significance. The Genetic Alliance, for instance, was a network formed about this time in which various advocacy groups joined forces with each other and with researchers to pool information and coordinate fund-raising, publicity, and research efforts. In one sense, voluntary associations like the Genetic Alliance and the FMSF had roots in a long tradition of self-help organizations, going back more than a century but particularly exemplified by Alcoholics Anonymous and its spinoffs. It was during the 1980s and early 1990s, however, that they aspired to become a force in scientific research and public policy. By 1990, voluntary organizations and networks were transforming how some scientific research was being pursued. The False Memory Syndrome Foundation looked very much like one of these new advocacy/research communities. But its dual character seemed to push the template to its limits.

The developing conversation about false memories quickly led to new research. After attending the Memory and Reality conference, and after hearing a talk at the American Psychological Association in August 1991 that blamed recovered memories of abuse on media articles and suggestive questions from psychotherapists,[37] Loftus began to brainstorm about the possibility of producing a false traumatic memory experimentally. How could she "scar" the brain "with something that never happened, creating a vivid but wholly imagined impression"?[38] Loftus put her question to her psychology class back at the University of Washington: "The trick was to design a study powerful enough to prove that it is possible to implant a false memory while also winning the approval of the university's Human Subjects Committee, which reviews proposed research projects to ensure that they will not be harmful to participants."[39]

She settled on the idea of "getting lost in a shopping mall," and she lost no time in trying it out on people who would later be called participants in a "pilot study."[40] She made her first informal experiment at a party, where she described her idea to a friend and asked if he thought she would be able to convince his eight-year-old daughter than she had gotten "lost in a shopping mall when she was five years old." The friend called his daughter over, introduced her to Loftus, and asked if she remembered getting lost in a mall when she was younger. After several rounds of assertions and suggestive questions, the girl stated that she had been scared when she got lost.[41] Later in the fall, Loftus offered the challenge to her cognitive psychology class for extra credit—two students came back with reports that they had been able to convince their younger siblings of the "lost in the mall" story.

The national media were already buzzing about Loftus's results before the "real" study had even begun.[42] The study itself was a bit different from the initial "pilots": each subject (in a group ranging in age from eighteen to fifty-three) took part in the study in collaboration with a relative. They were told that they were "participating in a study on childhood memories" and that the experiments were about "how and why people remembered some things and not others." The relatives provided a booklet containing brief accounts of three true incidents in the subjects' lives and, in addition, a plausible story about their getting lost in a shopping mall. The subjects were to fill out the booklets by writing what they could remember about each event, writing "I do not remember this" if they remembered nothing.[43] Of the twenty-four people who took part in the study, seven initially produced a false memory in response to the prompt about the mall. One of those changed his or her mind between the initial interview and a following one, leaving a success rate of 25 percent who remained at least partially convinced of the false memory by the end of the second interview.

Almost immediately, the study became a classic of sorts, even as it was attacked for exaggerating its results. To many people it seemed to demonstrate elegantly how easily a memory could be created out of whole cloth, which then became part of the furniture of a person's mind. It was quoted—and Loftus was asked to testify—in dozens of court cases in the 1990s involving long-ago memories of a crime, when the defense claimed that the events in question had not occurred. But there were also claims that Loftus had breached research ethics in various ways and that she had inflated the percentage of people in whom it had been possible to implant false memories.[44]

Loftus's experiment has already become something of a classic. It has been reproduced many times, with varying success. She has been accused of inflating her results, but some of the replications found higher rates of memory construction than she did. Hers was a very populist proposal, because anyone could try it.

One of the more recent replications was made by the online journal *Slate*, in a public trick that echoed Buckhout's 1970s phone-in experiment but was a bit more elegant. *Slate* altered several photographs in its files in dramatic ways, to create scenes that had never happened—for instance, it manufactured a picture of Barack Obama shaking hands with Iranian president Mahmoud Ahmadinejad. It mixed these fake photographs with real ones. Then *Slate* asked thousands of research subjects to describe their memories about each event. The false memory response rate ranged from 26 percent to 68 percent. When it revealed the hoax, *Slate* pointed to the power an entity like itself could wield if memory were so easily manipulated, declaring that it took as its model George Orwell's Ministry of Truth in *1984*:

> As soon as all the corrections which happened to be necessary in any particular number of *The Times* had been assembled and collated, that number would be reprinted, the original copy destroyed, and the corrected copy placed on the files in its stead. This process of continuous alteration was applied not only to newspapers, but to books, periodicals, pamphlets, posters, leaflets, films, soundtracks, cartoons, photographs—to every kind of literature or documentation which might conceivably hold any political or ideological significance. Day by day and almost minute by minute the past was brought up to date. In this way every prediction made by the Party could be shown by documentary evidence to have been correct, nor was any item of news, or any expression of opinion, which conflicted with the needs of the moment, ever allowed to remain on record. All history was a palimpsest, scraped clean and reinscribed exactly as often as was necessary. In no case would it have been possible, once the deed was done, to prove that any falsification had taken place.[45]

Slate "took the Ministry of Truth as our model" and made fiction a reality. Just as in Orwell's story, the changes it made to the historical record were not arbitrary. When it put Barack Obama and Ahmadinejad together it was hooking on to fringe—but, in a febrile political media, widely hinted—suspicions about Obama's loyalty and religion. Its other choices were similarly politically charged. For instance, it conjured up Senator Joe Lieberman voting to convict President Clinton at his impeachment trial, when in fact he had voted to acquit; only later did he acquire a reputation among Democrats as a turncoat. And it showed President Bush relaxing at his ranch with Roger Clemens during Hurricane Katrina, whereas in fact he had been in the White House on that day. The scenes related to suspicions, prejudices, or imagined possibilities already present in some people's minds, which *Slate* was about to "confirm" into memories. Some of Loftus's own work had suggested the same: she and colleagues had been able to show that memories for political events could be altered using doctored photographs.[46] All of this implied not only memory's changeability but its manipulability, on both an individual and a collective level.

Slate did not pretend to be able to erase "all records the way Orwell's ministry did," but new digital technologies gave it an unprecedented ability to alter them—and, as it was able to show, to change not only its own representations of the public record but also the autobiographical memories of its readers. Other claims about false memory in the early and mid-1990s were less novel. One was really a continuation of a debate of the 1980s, over the status of psychoanalysis. The 1980s saw what were sometimes called the "Freud wars," with accusations by some that Freud had behaved unethically (publicly disavowing his seduction hypothesis, for instance, while privately continuing to believe that his patients had actually been abused),[47] and claims by others that the whole structure of psychoanalytic theory was a confection promoted by a world-class confidence man. By the 1980s, psychoanalysis had lost quite a bit of the authority and glamour it had enjoyed in the middle years of the twentieth century; psychotherapists were struggling, as health maintenance organizations severely limited reimbursements for therapy sessions and the *Diagnostic and Statistical Manual* of the American Psychiatric Association moved away from psychoanalytic categories. The most visible attacks on psychoanalysis were, however, in the national newspapers and magazines, where a wholesale reconsideration of concepts like repression was taking place. The most eloquent of the attackers of psychoanalysis was the literary critic Frederick Crews of Berkeley, who had lambasted it as a spurious pseudoscience, a "shell game" promoted by the twentieth century's greatest intellectual charlatan. Crews now saw recovered memory as the latest and the most dangerous instantiation of Freudianism.[48]

Indeed, there was an air of nostalgia about some attacks—Orne's "demand characteristics," which had subsided to the role of a formulaic footnote in the methods section of papers, if it appeared there at all, now resurged as the term for how therapists "demanded" memory claims from their patients. Richard Ofshe and Margaret Singer, who had done their most memorable research in the 1970s and early 1980s, now found that the concept of a cult had a new freshness when applied to the survivor communities, and that their expertise in subtle forms of coercion was suddenly current and in demand.[49] False memory also found its way into standard psychology textbooks and other general writings on mind and memory published by highly respected psychologists like Daniel Schachter, Richard McNally, and others.

As skepticism waxed, psychotherapists found themselves challenged to show evidence that long-delayed recall of abuse was a phenomenon of real remembering, rather than of fantasy or suggestion, and that dissociation genuinely occurred in close correlation with abuse. The most serious difficulty was the demand that claims of sexual abuse be supported by separate corroborating evidence of the original events. After all, these kinds of crimes were highly private, and in cases of repressed memory they had taken place years earlier. A few remarkable instances did arise, such as the case of Loretta Woodbury, in which textual and other evidence surfaced from the long-distant past. It was exceedingly rare, however, to find corroborating evidence in a particular case so long after the event. It would be very hard to base a general defense of repressed memory on such scarce and chancy survivals. Researchers therefore turned to cases of abuse where such evidence was already known to exist, in order to examine the state of survivors' memories years after the fact. One study tracked the subsequent history of a child with a well-documented history of sexual abuse. She had not discussed the abuse with her family since the period when it occurred, and there were signs that she had forgotten it (she denied having been abused on various questionnaires). Then, at the age of nineteen, without the involvement of a therapist, her memories suddenly returned.[50] Similarly, a 1995 study looked at a group of young adults who had been removed from their homes a decade earlier. It found that out of nineteen women for whom there was evidence of serious sexual abuse, most recalled some of the documented events, but three had no memory and two of those described long blank periods around the times when the documented abuse had taken place.[51] Several other studies adopted a similar strategy of focusing on already documented cases of sexual abuse. For each claim made by the proponents of false memory, therefore, supporters of repressed memory could advance responses. But of course it was not the end of the conversation, because the false memory researchers scoured the references in these papers to find purportedly fatal flaws. Were these "documented" cases of abuse really so well documented? Repressed and false memory experts skirmished over retrospective diagnoses, poking at the raw

material for each others' claims. The balance of plausibility on the phenomenon at the heart of such much debate was never clear, stable, or settled—a
situation that continues to this day.

One of the most extreme and perhaps disquieting indications of the extent
to which protagonists were willing to press this conflict came with the case
of "Jane Doe." Doe was a woman who had been diagnosed as having suffered
sexual abuse when she was six, then appeared to have recovered the memory
of that abuse on videotape much later. When she was interviewed in 1984 by
psychotherapist Dr. David Corwin (at age six), she disclosed that she had
been abused by her mother. Corwin interviewed the girl again when she was
seventeen (again on videotape) and showed her the earlier taped interview.
During this interview, she initially seemed unable to recall any of the events
she had related eleven years earlier—but soon, during that very conversation, her memories returned. It was a sensational incident: Corwin not only
published the result but also played the videotape at several psychological
meetings.[52] But Loftus remained skeptical and resolved to research the case.
She was soon able to discover the real identity of "Jane Doe." Loftus then
tracked down her family and interviewed them, with a view to exposing what
she saw as a case of nonabuse. She published her account not in an academic
psychology venue but in the campaigning magazine *Skeptical Inquirer*.[53] Doe
herself was appalled at what she saw as a breach of research ethics and of her
privacy. She sued, in the process revealing her identity as Nicole Taus. After
years in the courts, the case was settled with a small financial concession
from Loftus's insurance company, but with immense other costs for Taus,
who was required to pay the fees for the other parties she had sued.[54] Despite
the energetic research of clinicians, by the middle 1990s, recovered memory
was in serious trouble. It had become a contested concept with rapidly wan-
ing plausibility. Hundreds of articles and television programs gave time to
"false memory," and each one further eroded recovered memory's credibility.
The reversal of fortune had been amazingly swift.

A Change of Policy

The mid-1990s were the moment of decision for recovered and false memory
disputes. Civil litigation and criminal charges peaked in the early and middle
1990s, providing a natural stage for the memory wars, which played out in
battles of expert witnesses across the country. After the Franklin conviction
was overturned and the Ramona third-party suit succeeded, it was clear that

such battles could be won—but it was not certain they would be. One of the most spectacular cases involving recovered memory, for instance, began with an accusation in 1996, when Paul Ingram, chairman of the Republican Party of Thurston County, Washington, and the chief civil deputy of the sheriff's department, was accused by his children of satanic ritual abuse (the children also accused other members of the sheriff's department). Ingram could remember no such thing but was persuaded to undertake therapy, where he began to remember a series of extreme and bizarre crimes that supported his children's claims. He confessed to these crimes and was charged accordingly. Ingram's defense attorney suspected that Ingram's memory was a construction, though Ingram himself did not. The attorney hired Ofshe to evaluate Ingram; Ofshe found him extremely suggestible and was able to make new suggestions to him that he accepted as his own memories. But by this time Ingram had been convicted. The false memory argument was made on appeal, but it was unsuccessful—Ingram served a large portion of a twenty-year sentence and was released only in 2003. Even as he served this time, his case came to stand for something hyperbolically wrong in the status of memory.[55]

The growing debate throughout the psychological disciplines created challenges for traditional professional organizations. The American Medical Association, the American Psychological Association, and the American Psychiatric Association all tried to frame a policy position that was responsive to their various constituents. Psychoanalytic psychotherapists and cognitive psychologists struggled intensely to find a compromise. In 1993 the American Medical Association declared that recovered memories were "of uncertain authenticity" and needed "external verification." The American Psychiatric Association warned that it was impossible to distinguish false memories from true ones.[56]

The American Psychological Association was the most agonized of these bodies. Its working group brought psychotherapists together with academic memory researchers. After three years of painful toil, they produced hundreds of pages of conflicting statements. There was almost nothing the two groups could agree on. Eventually they reached a compromise, but it pleased no one. They concluded that most people who were sexually abused as children remembered it. But one could also forget and later remember such abuse. And one could also construct "pseudomemories" of nonexistent events. The board of directors enlarged on the working group's statement by urging general caution. They warned that it was impossible to stipulate a specific set of symptoms in an adult that indicated childhood abuse. This meant that therapists should take a neutral position when patients described memories of childhood abuse. Potential clients, for their part, should be suspicious of overzealous therapists, such as someone who diagnosed childhood abuse early in therapy, in the absence of any such memories, or on the other hand who dismissed claims of sexual abuse "without exploration."[57]

In 1995, the American Psychological Association (APA) prepared a brochure on the issue, titled *Questions and Answers about Memories of Childhood Abuse*. It reported that some clinicians believed repression and recovered memories of traumatic childhood events were possible. But it also noted that "many researchers" felt "there is little or no empirical support for such a theory." It concluded that recovered memories were possible but rare; most abuse survivors "remember all or part of what happened to them." It seemed, not for the first or last time, that a major professional association was acting as a loosely coordinated network of distinct interests and commitments, the distinctions between them looming large at times of crisis. At a moment when the treatment and even the existence of such a major condition were at issue, the APA was institutionally impeded from reaching a definitive position. Failing to reach a clear and unified intellectual stance, the APA made do with a description of different kinds of diagnoses, such that every constituent group's stance was represented.

The APA committee's final report came out in February 1996. The panel was still deeply split. The members agreed that the memory wars should not undermine appropriate concern about the "pervasive" problem of childhood sexual abuse. They also agreed that victims remembered most sexual abuse. But beyond this they were at loggerheads. On the pivotal question of the evidential value of alleged long-buried memories, they produced a two-statement trade-off: memories of abuse that were forgotten for a long time *could* return to memory, but one could also form convincing "pseudomemories." After years of agonizing, this failed to resolve the crucial point of contention: some clinicians thought one could distinguish between the real and the pseudomemories, while others did not.

The Ends of False and Repressed Memory

In the late 1990s, repressed memory syndrome was in disgrace. Therapists struggled to defend themselves as parents sued them for damages. Survivors who had once found it comforting and energizing to speak out publicly now felt isolated. Indeed, while professional false memory proponents continued to be outspoken, it became all but impossible to connect with survivor communities. These communities—or at least individual survivors and probably informal networks of mutual support—still exist, but they are reclusive. It seems that the memory wars ended, and that false memory won.

As of 2011, the False Memory Syndrome Foundation is still in business, though barely, and is preparing to shut its doors. Pamela Freyd recounts that they were on the verge of closing a few years ago, but that was before the first prosecutions in the Catholic Church child abuse cases hit the headlines. Some of the earliest publicity about accused priests who have since been convicted of sexual abuse was picked up by the FMSF and included in its publications as exemplifying how much the repressed memory movement was like a witch hunt. The FMSF regarded aspects of the prosecutions—for example, the key role of a witness alleging recovered memories of abuses by the recently convicted Boston priest Paul Shanley, the most recent showdown over "false" versus "recovered" memory—as scandalous indications that false memories were still playing a destructive role in American society.[58] By this time, however, they no longer had the armies of volunteers and the fully staffed offices that would have allowed them to do battle in (and with) the media over the church child abuse scandal. They became primarily a resource for records should someone in the debate wish to consult what the FMSF had. In due course the foundation would become a repository for materials about the use of recovered memory in the church prosecutions themselves.

The first time I visited the foundation, in 2003, the place was eerily empty, with just Pamela Freyd and two others maintaining a daily presence. Not that this is necessarily a small number for an archive. The human beings on the premises were dwarfed by the paper and other recording media it contained. The foundation lived in an upscale three-story townhouse in which, until spring 2009, every scrap of documentation received during the memory wars was carefully preserved. Two or three floors of carefully indexed and cataloged materials covered about ten years of controversy. It was, and is, arguably the most comprehensive archive for any single high-profile public controversy in postwar America—at least in the public realm.[59]

Pamela Freyd continues to collect, index, organize, copy, and preserve these materials, and not just for immediate use within contemporary debates over memory. The prime raison d'être for the False Memory Syndrome Foundation is still to promote the reconstructive character of memory, but nowadays this has a subtly different focus. Its chief concern today is with therapists' own memories of the claims they made and the part they played in the memory controversies of the 1980s and early 1990s. Freyd believes the therapeutic communities will inevitably rewrite their own memories, archival and perhaps personal, to deny that they were among those who promoted the notion of recovered memory and associated effects like multiple personality disorder. Just as the FMSF declared that no *personal* memories could be trusted without external corroboration, so now it cautions that the psychological community will try to rewrite its *collective* memory. The only way to record what the FMSF considers the therapeutic community's most egregious excesses of self-delusion—to preserve a true collective memory—is therefore to meticulously retain all material records for future consultation.

In mid-2009 the False Memory Syndrome Foundation was preparing to shut its doors. Not that it would throw away a single scrap of paper: box after box was being shipped to Buffalo, New York, where a room had been set aside to house the materials in the Center for Inquiry. The Center for Inquiry is the home institution of the Committee for the Scientific Investigation of Claims of the Paranormal, or CSICOP. It is now known as the Committee for Skeptical Inquiry (CSI). Members of CSI have long supported the FMSF and played an active role in the memory wars. Now CSI sees the memory wars as a key episode in the chronicle of junk science and the FMSF as the exemplary instance of science allying with concerned citizens to police public culture.

The rise and proliferation of false memory syndrome indeed constitute a remarkable episode in the history of the human sciences, because of the systematic way "repressed memory" was challenged at both popular and academic levels and how quickly its credibility collapsed. The story leads us to a final, more speculative thought. I suspect that the memory wars became a major site where the public outside the world of academic "science studies" confronted claims of the social construction of scientific knowledge. The false memory proponents argued that recovered memory was a socially and culturally constructed entity, the result of a disastrous confluence of political trends, fashions in psychotherapy, and broader cultural developments in American society. The recovered memory supporters countered that the false memory advocates themselves were socially engineering a transformation in the plausibility of long-term and recovered memories. It seems ironic that in the science wars that briefly erupted in the late 1990s, the alleged slipperiness of the postmodernist position was associated with the recovered memory position, not with the false memory movement.[60] The science warriors associated recovered memory with woolly postmodernist doubt, and consequent moral anarchism, but aligned the FMSF stance with reasoned scientific skepticism. The irony is that the side of the debate that involved the most deliberate strategizing about how to achieve—through social and political action—the ascendancy of a particular intellectual stance over another, and indeed the recognition of a new fact, was the false memory community. In that sense false memory was the more socially constructed of the two entities—and therefore the one that emerged as comparatively robust, even real.

II

RELIVING AND REVISING THE PAST

Throughout the past century, researchers, therapists, litigators, policymakers, and surgeons have tried to find conditions where they could identify "authentic," pristine remnants of lived experience. If they could succeed, the implications were great. For instance, memory gaps could be interpreted as a sign of internal bad faith—an indication that one part of a person's mind is hiding information from consciousness or that one's original, whole self is maimed in some way and no longer capable of a true accounting of one's acts. The idea that in principle one is capable of remembering every scrap of experience—such as the shape of a crack in the ceiling in the bedroom where you slept when you were five—could become a sign of autobiographical integrity. This kind of claim has in fact been a theme of many of the projects discussed in previous chapters. It is no less present now.

These enterprises have never found unalloyed success, however, and many have failed altogether. Skeptics have derided as quixotic the very idea of seeking such records. They have used the concept as a foil against which to define a rival account of the true character of memory as constructive, messy, and changeable. But the projects had consequences, and the conflicts with such opponents even more so.

The persistence of memory projects seeking to recover the authentic past, and others rebutting this possibility, might suggest that there is a cyclical character to beliefs about memory in which we move between a stable model of "authenticity" and another of "reconstruction," turning from one to the other and back again in different circumstances. For instance, the forensic psychologists of today sometimes claim they are delivering the science that Hugo Münsterberg proposed; Frederic Bartlett is still cited by cognitive psychologists as launching the current "reconstructive" understanding of memory. But to see these two sides as fundamentally constant across the years in their commitments either to "authenticity" or to constructiveness would be to miss the point. Each new project mined earlier ones for raw materials, then used them to make something that fit its own situation and needs. The history of the memory sciences resembles in some respects the claims about memory made by one particular psychologist—Bartlett, whose mantra was that the fate, significance, and meaning of all psychic material was in the hands of future users.

Two current lines of research provide a glimpse of how questions of authenticity in memory are being handled today, and some of the ways these efforts appropriate and transform work from earlier in the century. Both address the neurophysiology of remembering. One is reminiscent of the midcentury surgical projects of Wilder Penfield and the supporters of Bridey Murphy. The other brings us into a bold new world of trauma neuropsychologists, who seek to develop ways of editing the emotional responses called up by particular events. Taken together, they provide a glimpse of how authenticity in memory is now being framed, and how the ambition to relive and remake our memory records continues to inform questions about memory and remembering. What follows is a snapshot of the way memory research is taking shape and its significance in the present day. One may expect these sensational new projects to give way in their turn, or even to be violently overturned in the manner of their predecessors earlier in the century.

Reliving in the Brain

In 2008, researchers in California performed an extraordinary experiment on a series of patients being prepared for brain surgery. These individuals suffered from a form of epilepsy that did not respond to drugs, and the surgery was an attempt to address it. To identify areas that were prone to seizures, the surgical team threaded tiny electrodes into the patients' brains, specifically the area in and around the hippocampus (because of the location of the seizure-prone tissue in these particular patients). This area of the brain had been associated since Penfield's time with the transformation of "present experience" into "future conscious recollections."[1] Although the electrodes were inserted for clinical purposes, they provided the opportunity for an unprecedented experiment that evoked some of the elements of Penfield's work, but with empirical results he could only have dreamed of.

Once the electrodes were in place and set to monitor the firing of a group of neurons, the researchers showed the patients a series of brief video clips, five to ten seconds each. Some of these videos were from situation comedies or talk shows; others were brief footage of animals, natural settings, and landmarks. While the patients watched the clips, the researchers recorded the patterns of activity of their neurons. In each case, certain neurons became highly active while others remained unresponsive. The researchers were able to graph individual neurons firing during a specific video sequence. One figure showed a subject's brain activity while viewing the *Oprah* show.

After a brief intermission, the researchers asked the patients to think back over what they had seen and describe whatever came to mind. They found that when a patient remembered a particular clip, the same pattern of neuron firings was repeated. As astonishing as this exact replication seemed to be, it was still more striking that the neuron activity seemed to begin a fraction of a second *before* the patient became aware of the memory. From this the researchers concluded that the neuron activity *produced* the experience of remembering—something that had been hypothesized much earlier but never empirically demonstrated. The identical pattern of neuron activity suggested that the brain "relived" past experiences during ordinary recall. As the researchers put it in their published report, the process by which memory—"a trace of things past"—emerges into consciousness was one of the "greatest mysteries of the human mind." It was this trace that they thought they had captured with their tiny electrodes.[2]

A similar idea had been proposed by Penfield in the 1950s and was much mocked by psychologists for its reductionism. The model of film that Penfield was so enamored of had led him to think of the neurons as preserving life for the "reliving" in a series of filmstrips, each neuron waiting its turn to fire in response to a perceptual experience. Film did play an important role in the new research, too. Its value lay in qualities Penfield would have appreciated: researchers could replay the source of an individual's experience over and over—not in the metaphorical sense of "replaying" (in memory) but literally, by simply showing the film clip, which could be relied on to be the same every time. It was through this reliability that researchers could be satisfied that the pattern of neuron firings responding to a particular clip was the same every time, both during spontaneous recall, when the patient was free to think of any clip, and when the memory was provoked by the researcher's choosing to play a particular clip. But film no longer had a part in how the researchers described the work of the memory neurons themselves. There was no "pathway" of neurons adduced to exist in the brain. Instead, the "experience" of, say, an interview with Tom Cruise on the *Oprah Winfrey Show* produced a bar graph of neuron activity levels. Still, the idea of a component of perceptual experience that could reside in the firing of a single neuron had been a key component of Penfield's claims, and in this research neuroscientists are asserting for the first time that they have recorded empirical evidence of it.

Other, somewhat less sensational research has supported similar conclusions. Rats trained to run through a maze in a certain alternating pattern of pathway choices had similar electrodes inserted in their brains. Researchers recorded the pattern of neuron firings for each maze route. Sometimes the rats would make a mistake and choose a route that was not part of the training pattern. Researchers found that they were able to predict when this would occur, because the neuron pattern appropriate to the mistake would be visible a fraction of a second before the rat's choice of route became clear.

The stir this research caused in the neurosciences was much greater than was registered in the general media. The findings were reported in major newspapers like the *New York Times*, but they did not provoke sustained reflections by, say, popular science bloggers. The significance of the work for our understanding of brain, mind, and memory was not immediately hailed. Perhaps the news failed to excite because, outside neurological and psychological research communities, it did not seem surprising. The idea that particular neurons were involved in "recording" particular memories is just what many, perhaps most, people would have expected had they given it much thought. This claim, "astounding" as some neuroscientists proclaimed it— and antithetical as some psychologists must have found it to their reconstructive understanding of memory—accords with a commonsense, populist understanding of memory records that has held remarkably constant over the course of a century.

Just as the UCLA research was appearing in print, however, another project was attracting far more public attention. Like the California research, this work involved the reliving of past experiences. It too made daring new claims about remembering and memory, but claims that were dramatically different in their implications from the neuron recording experiments and starkly opposed to the popular notion of faithful records. Instead of drawing a portrait of a perfectly reproducible "trace," it relied on the idea that every such reproduction provided a new opportunity for the "trace" to be altered.

Editing Memory

One of the most tenacious themes of twentieth-century memory research was the idea that people tormented by the memories of terrible experiences could benefit from remembering them, and from remembering them *better*. The assumption—broadly indebted to psychoanalysis—was that psychological records of traumatic events often failed to be fully "integrated" into conscious memories. As long as these records remained "dissociated," the sufferer was compelled to "relive" them instead of benignly remembering them. The more fully and appropriately one remembered terrible events, the more attenuated would be their emotional power.

But in the 1990s—a time when psychoanalytic assumptions were being challenged as never before—neuroscience researchers developed a new framework for thinking about remembering, forgetting, and the mind's record of past events. One result was a highly controversial new paradigm for treating traumatic memories. The problem with bad memories, these new researchers claimed, is not their complex and unresolved relation to one's sense of self, but the simple fact that they are unpleasant. These researchers defined emotional memory not in terms of repressed ideas, but by certain patterns of neuron action and the chemical changes they triggered. The next step was to change these patterns.

The story of this work is something of a departure in the sciences of memory, a tale that might seem an ironic finale after a century of efforts framed by the assertion and denial of memory's inviolability. The proposition here is that drugs can erase features of autobiographical memory, or at least "dampen" their associated emotions, and that this may in some circumstances be therapeutic. Yet the origins of the idea of *editing* memories by pharmaceutical means nevertheless lie in several strands of the memory research discussed in previous chapters. Memory erasure research thus offers a way of bringing us up to date on those trajectories too. Indeed, many of the threads of previous chapters come together as the context for this apparently unlikely story—a story that in the end seems to reverse the assumptions of many of the projects that supplied its initial ingredients.

Consolidation

It has long been understood that it takes some time for personal experiences to be fixed in the brain as long-term memories. But the term now associated with this idea—"consolidation"—dates from the 1960s. Neuroscience researchers at that time were seeking to develop a physiological account of learning. They sought to understand how and why memories took time to be formed in the mind and were not fixed as permanent records at the moment of the original experience. One of the answers to the "why" part of the question was developed in the 1960s and 1970s through research on emotion and memory. Several neuroscientists argued that a delay between perception and the formation of a permanent memory record gave the brain time to distinguish between important and unimportant memories and to make the more important memories stronger. How to define "importance" was, of course, its own question.

In the 1980s, several researchers said this process of modulation was controlled by the degree of emotional arousal attending an event. Important experiences, particularly traumatic ones, triggered the release of the adrenal stress hormones epinephrine and cortisol. It was this that enabled "the significance of an experience to regulate the strength of *memory* of the experience."[3] In the late 1980s and 1990s, psychiatrists and some neuroscientists saw the neurophysiology of stress hormones as providing the route to understanding the characteristic symptoms attributed to posttraumatic stress disorder. That is, they envisaged providing a physiological underpinning (and validation) for the central and controversial psychological explanation of PTSD: the notion that traumatic memories were stored differently from other memories, in a way that caused them to be repeated and relived over and over, long after the event itself, as if the event became "literally" part of the brain. During the peak years of PTSD research, neuroscience research on stress hormones did not deliver the robust connection to traumatic memory theory that PTSD researchers hoped for, although many findings seemed suggestive.[4] But although the neurophysiology of PTSD does not seem to have matured in the ways that some trauma therapists of the 1990s hoped, the neurosciences of emotion developed into a significant subfield of research, able to deliver results that captured the imagination of people far outside the field.

One of the earliest contributors to the idea that the significance of a memory determined how faithfully and vividly it was stored in the brain was none other than Robert B. Livingston, the neuroscientist whose "Now print!" hypothesis had supplied the underpinnings of flashbulb memories. Years later, when he was setting out his own arguments about consolidation, the Irvine neurobiologist James McGaugh referred to Livingston's work as helping to establish for him the relation between emotion and the tenacity and detail of certain memories. McGaugh also cited flashbulb memories as evidence for a positive correlation between emotional response to an event and information storage. And his own writings included, on occasion, anecdotes of a kind reminiscent of Livingston, Brown, and Kulik. He told, for instance, of a man who was only blocks from the Alfred P. Murrah Building in Oklahoma City on April 19, 1995, when the bomb set by Timothy McVeigh exploded. The man, McGaugh wrote, had described the experience as being "'engraved' in his brain, saying, 'I'll have that memory forever.'"[5]

McGaugh's research on emotional memory focused on such vivid, long-lasting memories of significant events because he was intrigued by how and why emotional events are remembered differently. He was convinced that they were indeed different from ordinary memories, but he had been unconvinced by Roger Brown and James Kulik's explanations, which related to the cognitive content of the memories themselves. Instead, he thought, the answer lay in stress hormone systems that were kicked into high gear when an experience generated a strong emotion. If memories were indeed modulated internally by hormones, this would allow the brain to adjust the consolidation of a given memory based on its significance, as marked by hormonal arousal. It had been accepted since the 1970s that stress hormone systems activated by emotional situations did indeed serve the "adaptive" needs of an organism.[6] In the 1990s his work extended this idea to argue that hormone systems played a role in the creation of long-term memories.

A contributing line of research suggested that although the amygdala was not the site where long-term memories were stored (and research had already established that records were actually *formed* in other parts of the brain, like the hippocampus), it mediated the influence of hormones on memory storage elsewhere in the brain.[7] McGaugh and others found that injecting chemicals that inhibited the synthesis of certain key proteins, or inhibited the action of receptors for those proteins, could affect the creation of memories.[8] They were able to enhance or inhibit the formation of long-term memories in rats by injecting into the amygdala drugs that enhanced or blocked its noradrenaline receptors. This suggested that the amygdala played a crucial regulatory role in the creation of long-term memories.[9]

There was an interesting twist to this research. The same procedures that could have an effect on the *initial* formation of a memory seemed also to affect memories *after* they were formed. Theories of consolidation had initially portrayed memories as permanently fixed in the brain, once the slow process of consolidation had been carried out. But even from the earliest days of consolidation theory, there were suggestions that memories might not be as permanent as this suggested. These indications came from various unusual, even extreme, neurological events. The psychiatric practice of electroconvulsive shock therapy (ECT), widespread in the 1950s and 1960s, for instance, was known to impair patients' memories. Among the wildly disparate claims about what ECT did to the brain were assertions that it removed painful memories, freeing traumatized subjects from their terrible pasts.[10] This did not amount to a claim that memories could change in the ordinary course of everyday life, only that it was possible to destroy them in extremis. But even this proposition was controversial, drawing objections that what had been destroyed was not the record itself but the ability to retrieve it.

These isolated claims led nowhere at the time, but they have recently made their way back into scientific references, as researchers have become curious about the genealogy of one of the hottest new theories of the past decade: "reconsolidation." The immediate origin of this concept lay in the 1990s study of emotional memory.[11] Researchers wanted to understand how the brain stored emotional memories, and how emotions affected the storage of other kinds of information. This research used Pavlovian fear conditioning on animals to create a memory of pain by coupling a neutral stimulus, like a particular sound, with a conditioning stimulus like an electric shock. After establishing the link between the tone and the shock, researchers would test the conditions under which the memories were stored, whether the memories could be strengthened or weakened, and whether they could be blocked or erased entirely. In one series of experiments, researchers injected certain chemicals into their subjects' brains. The goal was to enhance or inhibit the synthesis of proteins they thought were crucial to memory consolidation—to make a dramatic change either just before or just after a session in which the test animal was made to "retrieve" the memory (using the neutral stimulus alone). Later tests measured the strength of the response, in an attempt to see whether the drugs had affected the memory. For instance, Glenn Schafe and Joseph Ledoux of New York University reported evidence that the brain may alter a memory each time it is recalled. His research group trained lab rats to associate an electric shock with a particular sound. Whenever the tone sounded, they froze in fear. Ledoux and his colleagues gave the rats a break for a few days to allow the memories to become consolidated. In the meantime some rats received a drug inhibiting the creation of certain proteins in the amygdala that played a role in memory storage. Days later, they played the tone again. The drugged rats were no longer fearful, while the control group responded as strongly as ever.[12]

Researchers studying many species of animals found that chemicals that block the synthesis of certain proteins, or inhibit certain relevant receptors in the brain, can decrease the emotional content of a stored memory, at least under certain conditions. In animal research, this claim is based on the finding that the drugs in question can result in significantly weakened responses to the "neutral" stimulus that has been tied to the conditioning stimulus. Scientists had now established that hormones like cortisol had an effect on memory, not only at the time of the original experience but also later, when that memory was being retrieved. They seemed to affect not just the retrieval but the stored information itself. From this they concluded that memories had to be maintained by being consolidated anew—reconsolidated—after every retrieval, since it was possible to permanently alter them at these times.

The period when these animal researches were beginning to appear—the late 1990s—was opportune for anyone wishing to make a quick leap from animal research to the treatment of human beings. The decline of psychoanalytically oriented psychotherapy, the increasing financial pressure on therapists by health maintenance organizations, and the proliferation of a new generation of psychoactive drugs all made it easier to imagine a neuropsychiatric account of traumatic memory—one that might bring with it a psychopharmacological solution.

The claim that memories became volatile on recall raised the question whether it might be possible to intervene at this special time. A group of researchers at the University of California at Irvine found they could indeed influence people's emotional memories. They compared two groups of subjects, both of whom saw a series of slides accompanied by a story. One group heard a bland, innocuous narrative, but the other heard a more emotional one involving an injured child. The second group had much more detailed memories of the story and slides than the first group. A third group saw the same slides and heard the upsetting (second) story, but first they received a dose of propranolol. When this third group was tested, their memories of the details were no better than those of the group who heard the innocuous story. The conclusion the researchers drew was that "enhanced memory associated with emotional experiences involves activation of the beta-adrenergic system,"[13] and that blocking this action as the memories were being made could affect people's memories. Others were still more ambitious, trying the drug on sufferers of PTSD and finding that subjects who received the drug responded less strongly to "traumatic imagery" than those who did not receive it.

The old idea of "working through" painful memories suggested that the pain would become attenuated when the memories were better understood. Remembering these events over and over, and reflecting on them, was the healthiest course of action. This new approach asserts something more along the lines of "out of consciousness, out of mind." It follows suggestions, dating back to follow-ups of the original Brown and Kulik paper, that the more often a memory is recollected and described, the more vivid it becomes. This raises the question whether merely trying not to think about a particular memory could weaken it—something the old psychoanalytic paradigm would have rejected. Michael Anderson and John Gabrieli, psychology professors at the University of Oregon and Stanford, claim to have found support for this idea. When their test subjects avoided a specified memory, it gradually became weaker.[14] For neuroscientists pursuing "reconsolidation," this finding fit neatly with the idea that memories must be reconsolidated to remain strong. It was also part of a new account of PTSD. It proposed that PTSD worsens as you think about it, because each time you recall the painful events, the stress hormones that accompanied the painful memory are triggered and intensify that memory, making the emotions more vivid the next time it is recalled.

Erasable You

All of this has contributed to a new public conversation about the nature of the self, and about the relevance of our autobiographies to our relationships not only with friends and kin, but also with the broader community. The terms of reference of this conversation are of recent vintage. To get some perspective, it is worth comparing what people were saying about the relation between autobiography and selfhood in the slightly more distant past.

In the nineteenth century, character was commonly portrayed as something that was built up by daily experience and personal choices, through the memories and habits created by those everyday events. Character was the result of how one responded, moment by moment, to the challenges of daily life, because those responses built up a kind of internal machinery of habit. Character was defined, in a way, as an accretion of memory. The idea of "building character" meant striving to make appropriate choices because the hardware one created would become difficult or impossible to change later. One scholar of the present day who has expressed concerns about memory dampening has used a different analogy to describe the relation between memory and personality, but nevertheless one that describes personality as being made out of discrete memories. William B. Hurlbut, a consulting professor in human biology at Stanford University, wrote that "the pattern of our personality is like a Persian rug." It was built "one knot at a time, each woven into the others. There's a continuity to self, a sense that who we are is based upon solid, reliable experience. We build our whole interpretation and understanding of the world based upon that experience or on the accuracy of our memories."[15]

As recently as the 1990s, people who thought of themselves as survivors of terrible trauma often defined themselves in relation to what they remembered (or what they did not): they were survivors because they had survived certain defining events. Their character as mature adults came from "working through" these terrible memories. But there were also "survivors" who felt they had *not* truly survived their memories. They described a wounded self whose bad experiences stood in the way of personal realization. This latter convention involved the idea of a hidden, unimpaired self encumbered by adverse conditions. The idea is similar in some respects to the characterization used in the marketing of recent psychotropic drugs, especially Prozac. These drugs' enthusiasts sometimes declared that taking them allowed their "true" selves to emerge, often for the first time.[16] This suggests that there are at least two conflicting ways of thinking about memory and selfhood at stake in reflections about potential memory-altering agents: the survivor, using adverse events to become stronger, and the blocked self, emerging in a fully realized way with the aid of sophisticated new psychiatric techniques.

Philosophers, therapists, filmmakers, and bloggers have been quick to reflect on the implications of memory erasure. One of the best-known projects to explore the subject is the 2004 film *Eternal Sunshine of the Spotless Mind*, in which the main character attempts to have memories of his ex-girlfriend deleted from his mind pharmaceutically. The film embraces the new neuroscience of emotion, focusing on memories of feelings and the complex ways different kinds and parts of memories are stored in different places in the brain. Central to the plot is the idea that memories with different emotional associations are stored differently. Emotions associated with a past event can be stored independently of the "data" of the event itself, so that someone with amnesia about a particular event that took place, for instance, at a certain location could feel an echo of the emotion associated with that location in spite of not recalling what happened. The idea of reconsolidation is central not only to the basic plot (the lead man is made to remember his girlfriend so that memories of her can be systematically removed) but also to one particularly important element of the movie, in which some part of the consciousness of the main character becomes aware of the memory erasure as it is happening, and he changes his mind. Because he is anesthetized, he is unable to tell the operator to stop the procedure.

What follows evokes the idea of a Memory Palace—the medieval technique of committing things to memory by visualizing each piece of information as a room in a great palace. To recall a specific piece of information, one would imagine walking through the appropriate room. In the film, the lead character and his memories conspire together to save the memory records by stashing them in unlikely places on the palace grounds.[17]

The kind of cultural and business enterprise being envisioned here is indeed, as one acute reflection on the film puts it, "just the next logical step up from breast augmentation and Prozac."[18] And this is just the kind of image that a very different group of thinkers had in mind when they took on the possibilities of memory erasure. The President's Council on Bioethics, convened by President George W. Bush in his first term, thought the issue important enough to reflect on it alongside discussions of cloning and stem-cell research.[19] Editing memories could "disconnect people from reality or their true selves," the council warned. While it did not give a definition of "selfhood," it did give examples of how such techniques could warp us by "falsifying our perception and understanding of the world." The potential technique "risks making shameful acts seem less shameful, or terrible acts less terrible, than they really are."

The council suggested that altering memories could get in the way of the future, stronger self an individual might eventually become after grappling with the stresses of unpleasant memories. "The use of memory-blunters at the time of traumatic events could interfere with the normal psychic work and adaptive value of emotionally charged memory." It made an apparent reference to the literature on emotional memory, claiming that a "primary function" of the way the brain encoded emotional memories was "to make us remember important events longer and more vividly than trivial events." This function was valuable because emotional memories did indeed serve an "adaptive" purpose: we remember memories with emotional significance "longer and better" because they help us "learn, adapt, survive. Early hominids needed to know and remember that lions were dangerous. Modern children burn their fingers on a match and learn that fire hurts. We all learn to avoid bad things by remembering bad experiences."

Real Memory Therapies

The first speculative steps are now being taken in an attempt to develop techniques of what is being called "therapeutic forgetting." Military veterans suffering from PTSD are currently serving as subjects in research projects on using propranolol to mitigate the effects of wartime trauma. Some veterans' advocates criticize the project because they see it as a "metaphor" for how the "administration, Defense Department, and Veterans Affairs officials, not to mention many Americans, are approaching the problem of war trauma during the Iraq experience."[20] The argument is that terrible combat experiences are "part of a soldier's life" and are "embedded in our national psyche, too," and that these treatments reflect an illegitimate wish to forget the pain suffered by war veterans. Tara McKelvey, who researched veterans' attitudes to the research project, quoted one veteran as disapproving of the project on the grounds that "problems have to be dealt with." This comment came from a veteran who spends time "helping other veterans deal with their ghosts, and he gives talks to high school and college students about war." McKelvey's informant felt that the definition of who he was "comes from remembering the pain and dealing with it—not from trying to forget it." The assumption here is that treating the pain of war pharmacologically is equivalent to minimizing, discounting, disrespecting, and ultimately setting aside altogether the sacrifices made by veterans, and by society itself. People who objected to the possibility of altering emotional memories with drugs were concerned that this amounted to avoiding one's true problems instead of "dealing" with them. An artificial record of the individual past would by the same token contribute to a skewed collective memory of the costs of war.

In addition to the work with veterans, there have been pilot studies with civilians in emergency rooms. In 2002, psychiatrist Roger Pitman of Harvard took a group of thirty-one volunteers from the emergency rooms at Massachusetts General Hospital, all people who had suffered some traumatic event, and for ten days treated some with a placebo and the rest with propranolol. Those who received propranolol later had no stressful physical response to reminders of the original trauma, while almost half of the others did. The concerns of the President's Council suggest that those ER patients should have been worried about the possible legal implications of taking the drug. Could one claim to be as good a witness once one's memory had been altered by propranolol? And in a civil suit, could the defense argue that less harm had been done, since the plaintiff had avoided much of the emotional damage that an undrugged victim would have suffered? Attorneys did indeed ask about the implications for witness testimony, damages, and more generally, a devaluation of harm to victims of crime. One legal scholar framed this as a choice between protecting memory "authenticity" (a category he used with some skepticism) and "freedom of memory." Protecting "authenticity" could not be done without sacrificing our freedom to control our own minds, including our acts of recall.[21]

Authenticity and Identity

The anxiety provoked by the idea of "memory dampening" is intriguing. As we have seen, the concerns of the President's Council on Bioethics and others have been criticized as histrionic—far out of proportion to the practical feasibility of altering memories, now or in the near future.[22] But the mere possibility seems to have threatened an important convention for representing memory in relation to personal identity. I suggest that these worries draw their force from a deep-seated attachment to two related beliefs: first, that we are, in some ambiguous but important way, the accretion of our life experiences; and second, that those life experiences are perfectly preserved even if our ability to remember them is far from perfect. When Alzheimer's disease patients lose significant amounts of memory, dismayed friends often say that their very selves have crumbled or faded away and that in some literal way they are "no longer themselves."

The thought here is not that people believe their memories are perfect— far from it. Common understandings of memory centrally involve the idea that memories are unreliable, fickle, and capricious. But there is another belief about memory that has been articulated in various ways by many of the protagonists in these pages: that in some fundamental way, secreted within us are perfect records of past experiences, even if we might never access them consciously. This idea of a personal archive of the past, one that is somehow constitutive of selfhood, is a theme running throughout many of the projects chronicled in this book and doubtless has a far longer history.[23] It is most obvious, perhaps, in the story of Bridey Murphy. The importance of a continuous narrative past that could in principle be tapped at any point was implicit not only to the practical work of the memory therapists of World War II but to the "whole" self that they sought to help their patients create in themselves; and variations on this particular role for remembering were important to many projects related to traumatic memory for the rest of the twentieth century. Even in Penfield, the idea that an unconscious memory record serves consistently as a basis for recognition suggested that a highly detailed memory record is fundamental to personal identity. According to Penfield, we check the neural record of experience to know whether a new perception relates to something we have perceived before. If it does, we conclude that it is familiar and belongs to us.

A craving for neurophysiological "high fidelity" is a theme of virtually all the projects or controversies featured in this book—either to satisfy this longing or to demonstrate its futility. Münsterberg's projects were framed by the idea of the fidelity of memory, because the belief in this idea led witnesses and juries astray owing to their overconfidence in autobiographical memory or testimonial evidence.

A New Era of Remembering?

The concerns of the President's Council naturally raised the question of how to distinguish psychic pain from physical pain, since it was this very distinction that the council failed to discuss explicitly. The council did not set itself against the use of physical anesthesia, even though there are well-known cases where pain is thought to serve a useful purpose. Yet physical anesthetics have some relevant common ground with the prospective memory technologies, and there is a long history of resistance to the idea of physical anesthesia on grounds similar to some of the arguments being mounted here. Skeptics argued that a loss of sensation would disconnect sufferers from a valuable experience (as in childbirth) and from information they needed to have. Before the advent of anesthesia, techniques that seemed to involve an intentional suspension of sensation could trigger alarm on a scale that now seems almost inconceivable. For instance, in the 1830s, during disputes over whether mesmerism could create an altered state of mind in which an individual was entirely incapable of sensation, the editor of a major London medical journal urged his readers to consider such a thing impossible not merely because it was implausible but because it would be an immense moral affront and threat to one's personal integrity. "Consider the implications," he urged his readers: "the teeth could be pulled from one's head without one's knowledge." The point here is that we are no longer as disturbed by the idea that a tooth could be pulled painlessly, because we have a very different understanding of the relation between bodily pain and our sense of self. It seems that what is being proposed in memory-dampening research is part of a project that would implicitly reform our understanding of the relation between psychic pain and personal identity.

The 1840s and 1850s saw a dramatic shift in the status of physical pain, specifically in its relation to self-knowledge, self-control, and personal integrity. Anesthesia was initially upsetting because it challenged a convention for thinking about personal identity and self-control to which full sensory experience was central. Viewed from the era of anesthesia, this anxiety has come to look quaint or even inexplicable. In this era the suspension of sensation is taken to be radically distinct from how we understand ourselves. Could the same happen about remembering? The neurosciences of the present day are inaugurating a period of uneasy speculation about moral and identity issues that promises to be the most profound and far-reaching of all—at least until the next revolution in the memory sciences.

NOTES

Notes to Chapter One

1 "Crime of Fiend," *Chicago Chronicle*, January 14, 1906.
2 "Hopes to Escape Noose," *Chicago Chronicle*, March 6, 1906; "No Delay in Ivens Trial," *Chicago Chronicle*, March 7, 1906.
3 *Chicago Chronicle*, March 8, 1906.
4 *InterOcean*, March 20, 1906, 1; "Tells a New Story," *Chicago Chronicle*, March 20, 1906, 3.
5 "Healy to Refute Ivens," *Chicago Chronicle*, March 21, 1906, 3; "Seek an Alibi for Ivens," *Chicago Daily News*, March 15, 1906, 2.
6 Untitled article, *Chicago Daily News*, March 23, 1906, 1 and 2; "Say Ivens was Made to Confess Falsely," *Chicago Daily News*, March 15, 1906, 3.
7 "Normal Mind," *Journal of the American Medical Association (JAMA)*, February 25, 1896; "The Evidences of Insanity," *JAMA*, March 10, 1896; "Did He Sham on the Gallows? A Review of the Windrath Case," *JAMA*, June 27, 1896, 1274–75; J. Sanderson Christison, *The Tragedy of Chicago: A Study in Hypnotism, How an Innocent Young Man Was Hypnotised to the Gallows* (Chicago, 1906), 5.
8 Christison, *Tragedy of Chicago*, 9. He also cited an article by A. S. Walker in the *Edinburgh Medical and Surgical Journal*, 1898.
9 "Home by 9 PM Every Night," in Christison, *Tragedy of Chicago*, 11–13.
10 Herbert A. Parkyn, *Auto-suggestion: What It Is and How to Use It for Health, Happiness and Success* (Chicago: Science Press, 1906).
11 J. S. Christison, *The "Confessions" of Ivens: Did an Innocent Man Confess to a Great Crime?* (Chicago, 1906). J. S. Jastrow wrote on "subconscious" thinking; Frederick Peterson focused on the laboratory study of the emotions and on criminology; W. B. Pillsbury worked on attention; H. C. Warren studied consciousness; T. S. Clouston of Edinburgh encouraged people outside psychological fields to become more psychologically sophisticated; K. Eulenberger was an early "sexologist" with an interest in criminology; T. Claye Shaw of Britain studied consciousness and memory; Charles Richet of Paris had an interest in psychical research and

"hysterical" phenomena of amnesia and memory. On Christison's correspondence with William James, see William James, *Essays, Comments, and Reviews* (Princeton, NJ: Harvard University Press, 1987), 15:689–90.

12 As quoted in Christison, *Tragedy of Chicago*, 11.

13 This was by no means the accepted view, either of memory or of hypnosis, but was a new interpretation of the concept of "dissociation" developed by Pierre Janet.

14 "Psychology Put in Harness," *Chicago Tribune*, December 13, 1908; "Nothing to Fear in Hypnotism," *Chicago Tribune*, February 23, 1908.

15 Quoted in Hugo Münsterberg, *On the Witness Stand* (New York: McClure, 1908), 141–42 (I have not located the original source).

16 "Ivens to Seek Reprieve," *Chicago Chronicle*, June 21, 1906, 3; also "Ivens to Know Fate Soon," *Chicago Chronicle*, June 21, 1906, 3; "Ivens to Seek Reprieve," *Chicago Chronicle*, June 21, 1906, 3; "Ivens Writ Denied; Reprieve Is Sought," *Chicago Daily News*, June 21, 1906.

17 Münsterberg, *On the Witness Stand*, 142; "Ivens Hangs Today," *Chicago Chronicle*, June 25, 1906, 3; "Ivens Dies on Gibbet; He Fails to Confess," *Chicago Daily News*, June 22, 1906,

18 Matthew Hale, *Human Science and Social Order: Hugo Münsterberg and the Origins of Applied Psychology* (Philadelphia: Temple University Press, 1980), 113. For an account of Münsterberg that places his work in the larger context of science and the law, see Tal Golan, *Laws of Men and Laws of Nature: The History of Scientific Expert Testimony in England and America* (Cambridge, MA: Harvard University Press, 2004).

19 "'Hypnoscope' or Mirror Hypnotizer after Jules Bernard Luys"; see Arthur Mac-Donald, "Psycho-physical and Anthropometrical Instruments of Precision in the Laboratory of the Bureau of Education," in *Experimental Study of Children*, ed. Arthur MacDonald (Washington, DC: Government Printing Office, 1899), 1141–204, illustration on 1167). I learned about this device from Henning Schmidgen, "Münsterberg's Photoplays: Instruments and Models in His Laboratories at Freiburg and Harvard (1891–1893)," which provides a useful account of Münsterberg's instruments and laboratory. The Virtual Laboratory, http://vlp.mpiwg-berlin.mpg .de/references?id=art71.

20 Münsterberg, *On the Witness Stand*, 168.

21 Ibid., 169.

22 Ibid., 47.

23 "Klein Aims to Write Plays of the Hour," News clipping, Münsterberg Papers, Boston Public Library, 2499a (6), December 8, 1911. "Third Degree, His Latest Success, Is Second of a Series," *Pittsburgh Leader*, August 12, 1911; "Here Is Where Charles Klein Gets His Inspiration: He Finds Dramatic Plots in Newspaper Headlines," *Chicago Tribune*, February 10, 1910, B9.

24 "The Incubus on American Drama," *New York Times*, October 12, 1905. This article also reported that the play was doing well in New York's Hudson Theater.

25 Charles Klein, *The Third Degree: A Play in Four Acts* (New York, 1908), 19.

26 Ibid., 60.

27 "Coming Plays in Gotham by American Authors," *Chicago Tribune*, July 19, 1908, G2: "By persistent mental torture, prologued suggestion, and the influence of hyp-

notism, an innocent man may be forced into the confession of crime." Similarly, "Notes of Plays and Players," *Chicago Tribune*, October 25, 1908, H1.

28 Newspaper clippings, Münsterberg Papers, 2499a (6).

29 Newspaper clippings, Münsterberg Papers, 2499a (6); "Very Light Play at the Lyceum," *New York Times*, September 20, 1910, 11.

30 Review, *Moving Picture World*, May 17, 1919; similarly, *Wid's Daily*, May 8, 1919; Mordaunt Hall, "An Old Melodrama," *New York Times*, February 15, 1927.

31 Münsterberg, *On the Witness Stand*, 39–40.

32 Ibid., 143.

33 Ibid., 191–92.

34 See J. Anthony Lukas, *Big Trouble: A Murder in a Small Western Town Sets Off a Struggle for the Soul of America* (New York: Simon and Schuster, 1997), 588–99.

35 Herbert Nichols, "The Psychological Laboratory at Harvard," *McClure's Magazine* 1 (1893): 399–409.

36 Münsterberg, *On the Witness Stand*, 94.

37 "'I Can Tell If You're a Liar!' Professor with Strenuous Name Invents Machine That Will Make Him Famous," *New York Times*, September 15, 1907; "Precipitate Psychology," *New York Times* July 5, 1907, 6; "A Psychologist's Judicial Warning," *New York Times*, August 25, 1907, 16.

38 Münsterberg, *On the Witness Stand*, 110.

39 *Psychology and the Teacher* (1909); *Vocation and Learning* (1912); *Psychology and Industrial Efficiency* (1913); *Psychology, General and Applied* (1914, textbook); *Business Psychology* (1915 textbook for La Salle Extension University, Chicago).

40 Paramount's description was similar to later claims that radio would create a "university of the air."

41 Münsterberg had a planning meeting on December 22 with William W. Hodkinson. George R. Meeker, "editor of newspictures," wrote on January 6, 1916, that "we are in the midst of picturizing your scenarios." Münsterberg Papers. See also "Testing the Mind," a sample advertisement for *Moving Picture World*, Münsterberg Papers, 2443 and 2439.

42 Münsterberg to M. H. Swain, April 12, 1916, copy of outgoing letter, Münsterberg Papers, *2409; "Big Vital Thoughts Visualized on Screen," advertisement, *Evening Independent*, April 20, 1916.

43 Meeker to Münsterberg, May 16, 1916, Münsterberg Papers.

44 "Munsterberg Speaks at Paramount Reception, George Beban, and Other Notables at Exposition," *Motography* 15, no. 22 (May 27, 1916), 1215.

45 Wigmore to Münsterberg, July 28, 1907, Münsterberg Papers, *2244.

46 Wigmore to Münsterberg, 6 ALS, November 11, 1908, Münsterberg Papers *2244.

47 Münsterberg had claimed, according to Wigmore, that there were psychological publications that lawyers should have been familiar with, offering "exact methods of experimental psychology available and valuable for practical use in trials to measure testimonial certitude and to diagnose guilt." John H. Wigmore, "Professor Münsterberg and the Psychology of Testimony: Being a Report of the Case of *Cokestone v. Muensterberg*," *Illinois Law Review* 3 (1909): 399–445. On the effects of Wigmore's satire see Huntington Cairns, *Law and the Social Sciences* (New

York: Harcourt, Brace, 1935), 169, and James M. Doyle, *True Witness: Cops, Courts, Science, and the Battle against Misidentification* (New York: Palgrave Macmillan, 2005), 9–34.

48 Wigmore, "Professor Münsterberg," 415.

49 Ibid., 432.

50 John C. Burnham, *How Superstition Won and Science Lost: Popularizing Science and Health Care in the United States* (New Brunswick, NJ: Rutgers University Press, 1987), 11–12; Christopher Tourney, *Conjuring Science: Scientific Symbols and Cultural Meaning in American Life* (New Brunswick, NJ: Rutgers University Press, 2000), 8–9. On attitudes toward cultural change in the 1920s, see Lynn Dumenil, *The Modern Temper: American Culture and Society in the 1920s* (New York: Hill and Wang, 1995), 9.

51 See Tal Golan, *Laws of Men and Laws of Nature* (Cambridge, MA: Harvard University Press, 2004), chaps. 2 and 3 and, on Münsterberg, chap. 6.

52 Hugo Münsterberg, *Psychology and Life* (1899); *American Traits from the point of View of a German* (1901); *Die Amerikaner* (1904); *The Principles of Art Education* (1905); *Science and Idealism* (1906).

53 Wigmore to Münsterberg, ALS, December 17, 1912.

54 Wigmore to Münsterberg, December 2, 1910: "In that specific part which concerns the Stern methods, I lay emphasis on the jury's verdict as an essential part, even for scientific methods." Stern had laid down as "law" the idea that lapses of time after an event increased the likelihood both of forgetfulness and of false memory, but Wigmore cited other psychological authorities who contradicted him. See C. A. J. Coady, *Testimony: A Philosophical Study* (New York: Oxford University Press, 1995), 274.

Notes to Chapter Two

1 Otniel Y. Dror, "The Scientific Image of Emotion: Experience and Technologies of Inscription," *Configurations* 7, no. 3 (1999): 355–401.

2 Alex Owen, *The Darkened Room: Women, Power and Spiritualism in the Late Nineteenth Century* (London: Virago, 1989); Ruth Harris, *Murders and Madness: Medicine, Law, and Society in the Fin-de-Siècle* (Oxford: Clarendon Press, 1989); Alison Winter, "Harriet Martineau and the Reform of the Invalid in Victorian Britain," *Historical Journal* 38, no. 3 (1995): 597–617.

3 See Laura Otis, *Organic Memory: History and the Body in the Later Nineteenth and Early Twentieth Centuries* (Lincoln: University of Nebraska Press, 1994).

4 E.g., Alan Gauld, *History of Hypnotism* (Cambridge: Cambridge University Press, 1992), 450.

5 On the history of disputes about Freud's seduction hypothesis, see Christina Howard and Keith Tuffin, "Repression in Retrospect: Constructing History in the 'Memory Debate,'" *History of the Human Sciences* 15, no. 3 (2000): 75–93.

6 Margarete Sandelowski, *Pain, Pleasure, and American Childbirth: From the Twilight Sleep to the Read Method, 1914–1960* (Westport, CT: Greenwood Press, 1984). Scopolamine is now understood to act by blocking acetylcholine receptors in the parasympathetic and central nervous systems, an action that has drawn the interest

of researchers in the neurosciences of memory. See E. D. Caine, H. Weingartner, C. L. Ludlow, E. A. Cudahy, and S. Wehry, "Qualitative Analysis of Scopolamine-Induced Amnesia," *Psychopharmacology* 74, no. 1 (1981): 74–80.

7   R. E. House, "The Use of Scopolamine in Criminology," *Texas State Journal of Medicine* 18 (1922): 259.

8   Laurel Thatcher Ulrich, *A Midwife's Tale: The Life of Martha Ballard, Based on Her Diary, 1785–1812* (New York: Knopf, 1990), 147–66.

9   Virginia Leys, personal communication, May 23, 2001.

10 Russell Kelso Carter, *The Sleeping Car "Twilight," or Motherhood without Pain* (Boston: Chapple, 1915); Sandelowski, *Pain, Pleasure, and American Childbirth.*

11 See Ken Alder, *The Lie Detectors: The History of an American Obsession* (New York: Free Press, 2007); G. C. Bunn, "Spectacular Science: The Lie Detector's Ambivalent Powers," *History of Psychology* 10, no. 2 (2007): 156–78.

12 Charles Hanson Towne, *The Rise and Fall of Prohibition: The Human Side of What the Eighteenth Amendment and the Volstead Act Have Done to the United States* (New York: Macmillan, 1923), 130–31.

13 "Crime Wave Causes Daugherty to Act," *Dallas Morning News,* January 6, 1922; "Forum Will Hear of Lawlessness," *Dallas Morning News,* December 4, 1921; "The Lynching Record," *Dallas Morning News,* January 6, 1922; "No Quarter for Criminals," *Dallas Morning News,* January 11, 1922; "While Texas Sleeps," *Dallas Morning News,* December 16, 1921; "Big Crime Wave in United States," *Dallas Morning News,* February 10, 1922; "The Wave of Crime," *Dallas Morning News,* January 30, 1922.

14 "Big School for Detectives Is Latest Development," *Dallas Morning News,* December 18, 1921; "For the Want of Prosecution," *Dallas Morning News,* December 9, 1921; "Governor Urged to Save Negro: Appeal Made for Commutation of George Mckinley Grace," *Dallas Morning News,* January 1, 1922; "R. W. Harr Takes Stand to Deny Charge of Homicide," *Dallas Morning News,* February 1, 1922.

15 "Citizen's Duty regarding Enforcement of the Law," *Dallas Morning News,* January 1, 1922; "Solution of Prison Problem Discussed . . . Rests with People," *Dallas Morning News,* January 2, 1922.

16 The trial was requested by a Dallas district attorney, Maury Hughes, and witnessed by local officials.

17 C. Goddard, "How Science Solves Crime: 'Truth Serum,' or Scopolamine, in the Interrogation of Criminal Suspects," *Hygeia* 10 (1932): 981–83; "Prisoners Tell of Truth Drug Effect," *Galveston Daily News,* February 15, 1922.

18 "Doctors Discuss 'Truth Serum,'" *Dallas Morning News,* March 24, 1922.

19 Robert A. Hatcher, "Scopolamin-Morphin in Narcosis and in Childbirth," *JAMA* 54 (1910): 446; Bernard Kronig, "Scopolamine-Morphine Narcosis in Labour," *British Medical Journal* 2 (1908): 805–8.

20 As the city health officer for Ferris and assistant county physician to Ellis County, House would have come into contact with law enforcement officials. "Dr. R. E. House to Be Buried in Hometown," *Dallas Morning News,* July 16, 1930.

21 House, "Use of Scopolamine in Criminology"; "Negro Questioned under "Twilight Sleep" Freed of Charge of Murder," *Dallas Morning News,* February 21, 1922. In March, a medical convention addressed scopolamine: "Doctors Discuss 'Truth Serum,'" *Dallas Morning News,* March 19, 1922.

22 "Selections: Sodium Amytal in Criminal Investigations under Narcosis," *Medico-legal Journal* 40 (1923): 62–63; W. F. Lorenz, "Criminal Confessions under Narcosis," *Archives of Neurology and Psychiatry* 62 (1932): 82; "Scopolamin Attracts Interest," *Dallas Morning News*, February 2, 1922.

23 C. T. McCormick, "Deception Tests and the Law of Evidence," *California Law Review* 15, no. 6 (1926–27): 491–92; McCormick, "Deception Tests and the Law of Evidence," 328.

24 Ostensibly by the *Los Angeles Record* in 1922, according to sources of the 1940s and 1950s, but I have not been able to find an original source. The term "serum" was borrowed from the bodily "sera" that were used to make vaccines. Sera were thought to contain a subtle essence of the condition the vaccine was supposed to prevent. "Sera" also evoked fears that they might convey something of the animal donor's nature (see *Dr. Weiss' Brain Serum Injector* (Lubin Film Company, 1909–10).

25 "Prisoners Tell of Truth Drug Effect."

26 *Birmingham Age Herald*, January 7, 1924; *New York Times*, January 8, 1924.

27 "Cross-Examined under an Anaesthetic," *JAMA* 91, no. 25 (1928): 2006–7.

28 A. W. Herzog, "Scopolamine as a Lie Detector," *Medico-legal Journal* 40 (1923): 62–63; "San Quentin Officials Stand Aghast as Expert Conducts Demonstration," *San Francisco Chronicle*, June 27, 1923.

29 *Los Angeles Record*, June 30, 1923; *Sacramento Star*, June 30, 1923. For the New Orleans trip see *New Orleans Times-Picayune*, November 27, 1923. On House's experiments, see R. E. House, "Scopolamine-Apomorphine Amnesia," *Criminology*, no. 41 (1924): 162–68, and E. E. Free, "Strange New Crime Remedies," *Popular Science*, December 1926, 14–15.

30 R. E. House, "Why Scopolamine Should Be Made Legal," *Medico-legal Journal* 42 (1925): 138–48.

31 "'Truth Serum' Drug Tried on St. Louis Murder Suspect," *St. Louis Post-Dispatch*, June 3, 1924.

32 *St. Louis Post-Dispatch*, June 5, 1924. It seems Hulbert was not allowed to discuss the sessions in court.

33 *St. Louis Daily Globe-Democrat*, June 14, 1924.

34 "Results of Use of Truth Serum," *St. Louis Post-Dispatch*, June 13, 1924.

35 "*People v. Hudson*," in "Results of Use of Truth Serum," Missouri State Archives.

36 Andre A. Moenssens, "Narcoanalysis in Law Enforcement," *Journal of Criminal Law, Criminology, and Police Science* 52 (1961): 454; *State v. Hudson*, 31, 4 Mo. 599, 602, 289 SW 920, 921 (1926).

37 For some cultural context see Daniel Pick, *Svengali's Web: The Alien Enchanter in Modern Culture* (New Haven, CT: Yale University Press, 2000).

38 Alison Winter, *Mesmerized: Powers of Mind in Victorian Britain* (Chicago: University of Chicago Press, 1998), chap. 11.

39 House, "Use of Scopolamine in Criminology," 262.

40 "*People v. Hudson*," House deposition, in "Results of Use of Truth Serum," Missouri State Archives.

41 House, "Why Scopolamine Should Be Made Legal," 147.

42 *St. Louis Post-Dispatch*, June 9, 1924.

43 F. E. Inbau, "Scientific Evidence in Criminal Investigation," *Journal of the American Institute of Criminal Law and Criminology* 24 (1933–34): 1153.

44 J. M. Cattell, "Measurements of the Accuracy of Recollection," *Science* 2, no. 49 (December 6, 1895): 761–66; H. M. Cady, "On the Psychology of Testimony," *American Journal of Psychology* 35 (1924): 110–12; W. M. Marston, "Studies in Testimony," *Journal of Criminal Law and Criminology* 15 (1924): 5–31. G. M. Whipple, "The Observer as Reporter: A Survey of the 'Psychology of Testimony,'" *Psychological Bulletin* 6, no. 5 (1909): 153–70.

45 Henry Head, *Studies in Neurology* (London: Hodder and Stoughton, 1920), 605–6.

46 Hal Higdon, *Leopold and Loeb: The Crime of the Century* (Urbana: University of Illinois Press, 1999); Maureen McKernan, *The Crime and Trial of Leopold and Loeb* (London: Allen and Unwin, 1925).

47 E.g., John A. Larson, *Lying and Its Detection* (Chicago: University of Chicago Press, 1932), 99–121; Emmanuel Lavine, *The Third Degree: A Detailed and Appalling Exposé of Police Brutality* (New York: Vanguard Press, 1930), viii; Richard Sylvester, "A History of the 'Sweat Box' and 'Third Degree,'" *Proceedings of the Annual Convention of International Association of Chiefs of Police* 2 (1906–12): 25–33.

48 "Examination by Torture," *Outlook* 89 (May 30, 1908): 237–38, quoting the *New York Times*.

49 George W. Kirchway, "Does Scopolamine Make Criminals Tell the Truth, as Declared?" *Current Opinion* 75 (1923): 345.

50 "Medicine and the Law: Cross-Examination under Anesthetic," *Lancet* 106 (1928): 990–91.

51 R. E. House, "Use of Scopolamine in Criminology," *Texas State Journal of Medicine* 18 (1922): 259–63, at 259.

52 McCormick, "Deception Tests and the Law of Evidence," 398.

53 Herman Alder, "The Interests of Psychiatry in Criminal Procedure," *Representations to the American Bar Association* 16 (1922): 629–33, argued for a pivotal role for psychiatrists in interrogation.

54 John C. Goodwin, *Sidelights of Criminal Matters* (London: Hutchinson, 1923), 205; quoted in Larson, *Lying and Its Detection*, 100.

55 Inbau, "Scientific Evidence in Criminal Investigation"; F. E. Inbau, "Scientific Evidence in Criminal Cases," *Journal of Criminal Law and Criminology* 24, no. 6 (1934): 1154.

56 W. A. Dyche, "Science in the Detection of Crime," *Review of Reviews* 85 (1932): 52–54.

57 Inbau, "Scientific Evidence in Criminal Cases." See also Baker and F. E. Inbau, "Scientific Detection of Crime," *Minnesota Law Review* 177 (1933): 602, 623.

58 This is according to publicity photos in "Science Puts an End to 'Perfect Crime,'" *World's Fair Weekly*, June 17, 1933. On the staged tests, and for more on the involvement of the Scientific Crime Detection Laboratory (SCDL) in the Century of Progress Exposition, see Keith A. Barbara, "Scientific to the Last Degree: Northwestern's Scientific Crime Detection Laboratory and the Progressive Battle against Crime" (MA thesis, Johns Hopkins University, 1997), 50ff.

59 House's papers were listed on the syllabi, as was the published record of the Hudson case. R. E. House, "Laboratory Methods of Scientific Proof," in *Outline of*

Scientific Criminal Investigation (Ann Arbor: Northwestern School of Law, 1936), 43, Northwestern University Archives, Series 17/20 (Evanston, IL).

60 W. R. Kidd, *Police Interrogation* (New York: R. V. Basuino, 1940), 47.

61 Henry Morton Robinson, "Science Gets the Confession," *Forum* 93 (1935): 18.

Notes to Chapter Three

1 For a recent, trenchant intellectual history of psychological trauma, see Ruth Leys, *Trauma: A Genealogy* (Chicago: University of Chicago Press, 2000), chaps. 1, 3–4, and Ian Hacking, "Memory Sciences, Memory Politics," in *Tense Past: Cultural Essays in Trauma and Memory*, ed. Paul Antze and Michael Lambek (New York: Routledge, 1996).

2 Harold D. Palmer and Francis J. Braceland, "Six Years' Experience with Narcosis Therapy in Psychiatry," *American Journal of Psychiatry* 94 (1937): 49–50. Palmer and Braceland found that in depressed or catatonic patients narcosis therapy "renders the patient more accessible to psychotherapeutic endeavors and seems to enable the patient to reveal hidden sources of conflict."

3 Morris Herman, "The Use of Intravenous Sodium Amytal in Psychogenic Amnesic States," *Psychiatric Quarterly* 12, no. 7 (1938): 38–42. See also M. Abeles and P. Schilder, "Psychogenic Loss of Personal Identity," *Archives of Neurology and Psychology* 34 (1935): 587; F. H. Leavitt, "The Etiology of Temporary Amnesia," *American Journal of Psychiatry* 91 (1935): 1079; M. W. Thorner, "The Psychopharmacology of Sodium Amytal," *Journal of Nervous and Mental Disease* 81 (1935): 161.

4 J. S. Horsley, "Narco-analysis," *Journal of Mental Science* 82 (1936): 417. He was a psychoanalyst, but his immediate access to the inner recesses of patients' minds could have reminded readers of his surgical background (his father was the renowned surgeon Victor Horsley).

5 W. Sargant and E. Slater, "Acute War Neuroses," *Lancet* 2 (1940): 1–2. On Sargant see Leys, *Trauma*, 190–228, and on wartime neurosis more generally, see Ben Shephard, *A War of Nerves: Soldiers and Psychiatrists in the Twentieth Century* (London: Jonathan Cape, 2000).

6 W. Kubie and S. Margolin, "The Therapeutic Role of Drugs in the Process of Repression, Dissociation, and Synthesis," *Psychosomatic Medicine* 7 (1945): 147–48.

7 William Sargant and Eliot Slater, *An Introduction to Physical Methods of Treatment in Psychiatry* (Edinburgh: Livingstone, 1944), 90. This text also recounts how Sargant's own staff routinely dosed themselves with sodium amytal to get through the harrowing months of the Blitz in 1940.

8 William Sargant and Eliot Slater, "Amnesic Syndromes in War," *Proceedings of the Royal Society of Medicine* 34 (1941): 763.

9 Shephard, *War of Nerves*.

10 Michael Clarke, Wellcome Institute for the History of Medicine, personal communication.

11 Patient records, Sutton Emergency Hospital, PP/WWS, Wellcome Library for the History of Medicine. These records are unpaginated and sorted alphabetically by patient name. For confidentiality, no precise reference is given here.

12 See Leys, *Trauma*, 200–203, for a detailed and thoughtful account of Sargant's shift to ether.

13 J. F. Wilde, "Narco-analysis in Treatment of War Neuroses," *British Medical Journal* 2 (July 4, 1942): 4; *Medical Press and Circular* 205 (February 12, 1941): 7; J. S. Horsley, *Narco-analysis, a New Short-Cut Technique in Psychotherapy* (London: Oxford University Press, 1943), paraphrasing the American neurologist and psychiatrist Smith Ely Jelliffe on 39; W. Sargant, "Physical Treatment of Acute War Neuroses: Some Clinical Observations," *British Medical Journal* 2 (1942): 574.

14 J. L. Clegg, "Narco-analysis in Treatment of War Neuroses," *British Medical Journal* 2 (1942): 140–41.

15 Gerold Frank, *The Boston Strangler* (New York: New American Library, 1966), 266–67.

16 Memory drugs were to be used along with electroshock therapy, prolonged sleep, sensory deprivation, and, after significant memory loss, the use of taped messages. Harvey Weinstein, *A Father, a Son and the CIA* (Toronto: James Lorimer, 1988), 138; Gordon Thomas, *Mindfield* (Denver: Mentor Books, 2001), 273–74; On Ewan Cameron see John Marks, *The Search for the "Manchurian Candidate": The CIA and Mind Control* (New York: Times Books, 1979), 144.

17 E.g., M. S. Guttmacherm, "Consultation Services (Mental Hygiene Clinic)," from "Consultation—Mental Hygiene," National Archives and Records Administration (NARA), Neuropsychiatry. See also Roy W. Menninger and John C. Nemiah, eds., *American Psychiatry after World War II (1944–1994)* (Washington, DC: American Psychiatric Press, 2000), 8.

18 Joseph McNinch, Abstract of John W. Appel's "Cause and Prevention of Psychiatric Disorders in the U.S. Army in World War II," NARA Historical MSS, Neuropsychiatry. McNinch reported that psychiatric disorders were a major drain on manpower—a million admissions to hospitals and half a million psychiatric discharges; the budget allowed one psychiatrist per hospital of 250 beds or more but offered no psychiatric training for doctors, medical technicians, or unit commanders. Norman Q. Brill, "[Neuropsychiatry]," 1; S. A. Challman, "History of Neuropsychiatry in the Southwest Pacific Area," 13, 14: "Fully half of the psychiatrists on duty in the theatre in 1942 were not professionally qualified for work without supervision." Norman Q. Brill, "Neuropsychiatry—Hospitalization in the Zone of Interior," draft #2. Two dedicated stateside psychoneurosis hospitals were created in 1942–43 (Darnall Hospital in Danville, Kentucky, and Mason General Hospital on Long Island); Brill, "[Neuropsychiatry]." All from NARA, Historical MSS—Psychoneuroses.

19 Joseph McNinch, Abstract of John W. Appel's "Cause and Prevention of Psychiatric Disorders in the U.S. Army in World War II." MicNinch cited a memo from Menninger to Major General Roy E. Porter: "Mental Health of Military Personnel," December 18, 1943, NARA, file SPMCB.

20 Brill, "Neuropsychiatry—Hospitalization in the Zone of Interior."

21 Brill, "[Neuropsychiatry]," 1, 10: "In the latter part of 1940, there were very few regular army medical officers who were trained in psychiatry. . . . Great improvements came with the creation of psychiatric centers where patients could receive . . . 'Narcosynthesis' by Grinker and Spiegel, which had gained great popularity overseas."

22 Mark K. Wells, *Courage and Air Warfare: The Allied Air Crew Experience in the Second World War* (London: Frank Cass, 1995); Shephard, *War of Nerves*, 286–88.

23 Edgar Jones and Simon Wessely, "'Forward Psychiatry' in the Military: Its Origins and Effectiveness," *Journal of Traumatic Stress* 16, no. 4 (2004): 411–19; Menninger and Nemiah, *American Psychiatry after World War II*, 8–9.

24 Menninger and Nemiah, *American Psychiatry after World War II*.

25 Roy R. Grinker and John P. Spiegel, *War Neuroses* (Philadelphia: Blakiston, 1945), 78; Roy R. Grinker and John P. Spiegel, *Men Under Stress* (Philadelphia: Blakiston, 1945), 170.

26 Grinker and Spiegel, *Men Under Stress*, 172–73. Similarly, Lloyd Thompson [Army Report, 312th station N.P. hospital], 109.

27 Grinker and Spiegel, *Men Under Stress*, 173.

28 Ibid.

29 Grinker and Spiegel, *War Neuroses*, 78.

30 Grinker and Spiegel, *Men Under Stress*, 174.

31 Grinker and Spiegel, *War Neuroses*, 78. Indeed, this was why Grinker preferred Pentothal over Amytal; with the latter drug, a period of prolonged sleep followed the trance.

32 The arguments were a bit conflicted; Grinker and Spiegel wrote that narcosynthesis was not suitable for use in a forward location—only for "treatment in base areas." But they also urged that treatment should be attempted as quickly as possible, with "sedation" at the forward areas.

33 "It was a regular medical procedure without mysticism and the technique was more readily accepted and learned by medical officers." Lloyd J. Thompson, "History of the Neuropsychiatric Services in the European Theater of Operations, 107, NARA, Historical MSS—Neuropsychiatry.

34 Ibid., chap. 10, "Treatment of Neuropsychiatric Conditions."

35 Ibid., 94.

36 "NP Combat Casualties," *Hospital Corps Quarterly* (supplement to *U.S. Naval Medicine Bulletin*), September 1945, 23–28, 24–25, 28. See also A. Kardiner, "The Neuroses of War," *War Medicine* 1 (1941): 220: "I . . . say that every one exposed to war has a mild form of traumatic war neurosis."

37 After the experience "had been relived in detail," Fabing used a drug called Coramine to awaken the patient, while the doctor reviewed the revived memory to retain it in the patient's consciousness. A special wardman "who had participated in the drama" continued this review work for several hours. On Fabing see Leys, *Trauma*, 215–17.

38 Thompson, "History," 109. See also Leon Altman, Louis Pillersdorf, and Alexander Ross, "Neurosis in Soldiers: Therapeutic Barriers," *War Medicine* 2 (July 1942): 551–60, on the relationship between doctor and patient.

39 M. B. Wright, "Treatment of Psychologic Casualties during War," *British Medical Journal* 2 (1939): 615.

40 "Under hypnosis, the soldier begins to re-enact his battle experiences as if they were in the present. . . . An essential aspect of the therapeutic procedure was the subsequent discussion in detail with the patient of the material which had been brought out under hypnosis. The acceptance and mastery by the ego of the traumatic experiences

were a fundamental part of his therapy." M. Ralph Kaufman, "A Psychiatric Treatment Program in Combat," NARA, Historical MSS—Neuropsychiatry.

41 Henry Beecher, "Anesthesia's Second Power: Probing the Mind," *Science* 105, no. 2720 (1947): 164–66.

42 Lesley Brill, *John Huston's Filmmaking* (New York: Cambridge University Press, 1997), 111.

43 John Huston, *An Open Book* (New York: Knopf, 1980), 125, as quoted in Michael Shortland, "Screen Memories: History of Psychiatry and Psychoanalysis on Film," *British Journal for the History of Science* 20 (1987): 431.

44 Lawrence Grobel, *The Hustons* (New York: Scribner's, 1989), 272.

45 James Agee in the *Nation*, May 11, 1946. Cited in Brill, *Huston's Filmmaking*, 112. See also Grobel, *Hustons*, 273.

46 Huston, *Open Book*; Brill, *Huston's Filmmaking*, 112.

47 A few years later, the army made a substitute for *Let There Be Light*, called *Shades of Gray* (Signal Corps, 1948). It was more didactic, with no shots of patients in distress. The concern about such images could have been intensified by anxiety about the depleted status of the armed forces just after the war. See Russell Frank Weigley, *History of the United States Army* (Bloomington: Indiana University Press, 1984), 486–500.

48 C. Lambert and W. L. Rees, "Intravenous Barbiturates in the Treatment of Hysteria," *British Medical Journal* 2 (1944): 73.

49 W. G. Lennox, "Real and Feigned Amnesia: Scientific Proof and Relations to Law and Medicine," *American Journal of Psychiatry* 99 (March 1943): 732–45; D. P. Morris, "Intravenous Barbiturates: An Aid in the Diagnosis and Treatment of Conversion Hysteria and Malingering," *Military Surgeon* 96 (July 1945): 509–13; H. Oliver, "Medicolegal Value of Psychosomatic Narcoanalysis in Its Military Application," *Annales Médecine Légale* 27 (August–September 1947): 154; E. L. Lipton, "The Amytal Interview: A Review," *American Practical Digest of Treatment* 1 (1950): 148–63. "Evidence Obtained under the Influence of Drugs or Hypnosis," *Medical-Legal and Criminological Review* 10 (1942): 66–67.

50 A. O. Ludwig, "Clinical Features and Diagnosis of Malingering in Military Personnel," *War Medicine* 5 (1944): 378–82. On the earlier history of similar issues, see Michael R. Trimble, *Post-traumatic Neurosis from Railway Spine to the Whiplash* (New York: Wiley, 1981).

51 E.g., D. P. Morris, "Intravenous Barbiturates: An Aid in the Diagnosis and Treatment of Conversion Hysteria and Malingering," *Military Surgery* 96 (1945): 509–13; C. P. Adatto, "Observations on Criminal Patients during Narcoanalysis," *Archives of Neurology and Psychiatry* 62 (1949): 88.

52 John Brunner, *OSS Weapons* (Washington, DC: Office of Strategic Services, 1943–44), 91.

53 Bernard Rubin, personal communication, 2009.

54 According to Dr. Bernard Rubin, a colleague of his, Grinker's purpose was to try to get at dissociated memory and by this means reconstruct the story of the crime (personal communication).

55 Years later, Grinker supplied Heirens with a statement that he had not confessed

during the session. I am grateful to Bernard Rubin for showing me his copy of this document.

56 United States of America ex rel. *William Heirens, Petitioner-Appellant v. Frank J. Pate, Respondent-Appellee*, 405 F.2d449.

57 There have been conflicting accounts disputing Heirens's guilt, several appeals, and a recent request for clemency. See D. O. Cauldwell, *William Heirens: Notorious Sex Maniac* (Chicago, 1948); Dolores Kennedy, *William Heirens: His Day in Court/Did an Innocent Man Confess to Three Grisly Murders?* (Chicago: Bonus Books, 1991); Lucy Freeman, *"Before I Kill More . . .": The William Heirens Story, an Account and a Quest* (New York: Crown, 1955); A&E documentary, *American Justice: Who Is the Lipstick Killer?* (DVD released 2008).

Notes to Chapter Four

1 Wilder Penfield, "The Cerebral Cortex in Man: 1. The Cerebral Cortex and Consciousness," *Archives of Neurology and Psychiatry* 40, no. 3 (1938): 431.

2 John Hughlings Jackson, "On Right or Left-Sided Spasms at the Onset of Epileptic Paroxysms, and on Crude Sensation Warnings, and Elaborate Mental States," *Brain* 3 (1880–81): 192–206.

3 Wilder Penfield, "The Mechanism of Memory," *Transactions of the American Clinical and Climatological Association* 62 (1950): 2.

4 Ibid., 4–5.

5 Ibid., 9.

6 Ibid., 2, 5–6.

7 Ibid., 7.

8 E.g., Wilder Penfield, "Some Problems of Wartime Neurology," *Archives of Neurology and Psychiatry* 47 (May 1942): 839–41; Wilder Penfield, "Neurosurgery in the War Period," in *Proceedings of the Second Annual Meeting, American-Soviet Medical Society*, Philadelphia, December 15, 1945.

9 William Feindel, personal communication, 2005.

10 Karl Lashley, "In Search of the Engram," *Society of Experimental Biology Symposium* 4 (1950): 455.

11 "Any two cells or systems of cells that are repeatedly active at the same time will tend to become 'associated,' so that activity in one facilitates activity in the other." Donald O. Hebb, *The Organization of Behavior: A Neuropsychological Theory* (New York: Wiley, 1949), xix.

12 Lashley, "In Search of the Engram."

13 Brenda Milner, personal communication.

14 W. B. Scoville, "Limbic Lobe in Man," *Journal of Neurosurgery* 11, no. 1 (1954): 64–66.

15 W. B. Scoville and B. Milner, "Loss of Recent Memory after Bilateral Hippocampal Lesions," *Journal of Neurology, Neurosurgery, and Psychiatry* 20 (1957): 11–21; S. Corkin, "What's New with the Amnesic Patient H.M.?" *Nature Reviews Neuroscience* 3 (2002): 153–60.

16 http://thebrainobservatory.ucsd.edu/hm_live.php.

17 William Feindel and Wilder Penfield, "Localization of Discharge in Temporal Lobe Automatism," *AMA Archives of Neurology and Psychiatry* 72, no. 5 (1954): 605–30; John Eccles and William Feindel, "Wilder Graves Penfield, 1891–1976," *Biographical Memoirs of Fellows of the Royal Society* 24 (1978): 481. On automatism see Wilder Penfield, "Epileptic Automatism and the Centrencephalic Integrating System," *Proceedings of the Association for Research in Nervous and Mental Diseases* 30 (December 15–16, 1950): 513–28; Wilder Penfield, "Some Observations on Amnesia," offprint of *American Journal of Psychiatry* 110 (1954): 11.

18 Penfield, "Mechanism of Memory," 14.

19 Ibid., 7–8.

20 Wilder Penfield, "Memory Mechanisms," *AMA Archives of Neurology and Psychiatry* 67 (1952): 178–91.

21 "A Chance for Elizabeth," *Time*, November 29, 1948, 71–72.

22 Lawrence S. Kubie, "Some Implications for Psychoanalysis of Modern Concepts of the Organization of the Brain," *Psychoanalytic Quarterly* 22 (1953): 29.

23 Brenda Milner, August 2007, personal communication.

24 Kubie, "Some Implications," 32–33.

25 Ibid., 46–47.

26 Ibid.

27 Penfield, "The Permanent Record of the Stream of Consciousness," *Proceedings of the Fourteenth International Congress of Psychology*, Montreal, June 1954, 67–68.

28 Ibid.

29 Wilder Penfield, "Consciousness, Memory, and Man's Conditioned Reflexes," in *On the Biology of Learning*, ed. Karl H. Pribram (New York: Harcourt, Brace, and World, 1969), 165.

30 See Patricia Zimmerman, *Reel Families: A Social History of Amateur Film* (Bloomington: University of Indiana Press, 1995), chaps. 4–5.

31 "Trivial Treasures," editorial, *Montreal Gazette*, February 25, 1955. Similarly, another reporter got tangled up in a mixed metaphor: There was "something like a sound movie film of your whole life in your brain." If a surgeon knew "where to set the needle," he could play back any moment. Ed Hadley, "Movie Film in Brain: Penfield Reveals Amazing Discovery," *Montreal Star*, February 14, 1957.

32 "The Brain as Tape Recorder," *Time*, December 23, 1957, 36.

33 Harry Everett Smith, *Think of the Self Speaking: Harry Smith—Selected Interviews*, ed. Rani Singh (Seattle: Elbow/Cityful Press, 1999), 57–58. I am grateful to Tom Gunning for calling my attention to Smith and his interest in Penfield.

34 On Smith's use of Penfield, and on *Heaven and Earth Magic*, see P. Adams Sitney, *Visionary Film: The American Avant-Garde, 1943–2000*, 3rd ed. (New York: Oxford University Press, 2002), 255ff. See also Judith Switzer, "The Spatial Strategies of Harry Smith's *Heaven and Earth Magic*," *Film Reader* 3 (1978): 207–22. In contrast to Penfield's representation of both memory and film, Smith's films were not about conveying reality, but about the hallucinatory experience of some scrap of reality turned into something *sur*real. See also Jamie Sexton, "Alchemical Transformations: The Abstract Visions of Harry Smith," *Senses of Cinema*, no. 36 (July 2005), in www.sensesofcinema.com/2002/harry_smith.html.

35 Rani Singh, "Last Day in the Life of Harry Smith," *Film Culture* 76 (June 1992): 20; Mary Hill, "Harry Smith Interviewed," *Film Culture* 76 (June 1992): 1–7; Sitney, *Visionary Film*, 246.

36 Noël Carroll, "Mind, Medium and Metaphor in Harry Smith's *Heaven and Earth Magic,*" *Film Quarterly* 31 (1977–78): 2, 37; Sitney, *Visionary Film*, 261.

37 Lee Strasberg, *A Dream of Passion: The Development of the Method* (New York: Penguin 1987), 113–14. See also Paul Kuritz, *Playing: An Introduction to Acting* (Englewood Cliffs, NJ: Prentice-Hall, 1982), 37–38.

38 E.g., Colin Blakemore, *Mechanics of the Mind* (London: Cambridge University Press, 1997), 49–51; Wilder Penfield, *The Mystery of the Mind* (Princeton, NJ: Princeton University Press, 1975).

39 Brenda Milner, "Memory Mechanisms," *CMA Journal* 116 (June 18, 1977): 1374–76.

40 Elizabeth F. Loftus, *Eyewitness Testimony*, rev. ed. (Cambridge, MA: Harvard University Press, 1996), 116ff.

Notes to Chapter Five

1 R. M. Thomas, "Morey Bernstein, Proponent of Bridey Murphy, Dies at 79," *New York Times*, April 11, 1999, 296.

2 T. E. Mullaney, "Scrapping for Scrap; Waste Metal Is Basic in Production Today," *New York Times*, October 7, 1951, 207.

3 J. B. Rhine, *New Frontiers of the Mind: The Story of the Duke Experiments* (New York: Farrar and Rinehart, 1937).

4 Gina Cerminara, *Many Mansions: The Edgar Cayce Story on Reincarnation* (New York: Sloane, 1950); Edgar Cayce, *Auras: An Essay on the Meaning of Colors* (Virginia Beach, VA: ARE Press, 1945).

5 For the current location and management of the Edgar Cayce archives, see http://www.edgarcayce.org/edgarcaycefoundation/.

6 *Denver Post*, September 19, 1954, 12.

7 Ibid.

8 The dates of these sessions were December 18, 1952; January 22, 1953; July 27, 1953; and October 1, 1953, respectively.

9 Tighe to Bernstein, April 10, 1953; Bernstein papers (BP), Pueblo Historical Society.

10 W. J. Barker, "The Strange Search for Bridey Murphy, *Denver Post*, September 12, 1954, 7–8. Part 2 of this article appeared in *Denver Post*, September 19, 1954, 12.

11 Tighe to Bernstein, September 21, 1954, BP.

12 Bernstein to Tighe, May 10 1955, BP.

13 Bernstein to Tighe, October 22, 1955, BP.

14 Bernstein to Tighe, December 12, 1955, BP.

15 Fred Grunfeld, "Murphism on LP," *Saturday Review*, April 28, 1956.

16 "Morey Bernstein Dies; Wrote 1950s Bestseller," *Washington Post*, April 8, 1999.

17 Louella O. Parsons, "Quartet to Play 'Bridey Murphy,'" *Chicago American*, May 21, 1956. Wright had a well-established career since her first film, *The Little Foxes* (1941),

then more recently for a role in the Hitchcock film *Shadow of a Doubt* and two others, *The Best Years of Our Lives* (1946) and *The Men* (1950).

18 Thomas, "Morey Bernstein," 296.

19 Marie Torre, "TV-Radio Today," *New York Herald Tribune*, March 15 1956; "NBC Plans 'Bridey' Probe," *Daily News*, April 3, 1956.

20 Grunfeld, "Murphism on LP."

21 [William Buckley], "Belfast in Colorado," *National Review*, March 7, 1956, 1.

22 Henry Atwell, "The Myth of Bridey Murphy," *Catholic Courier-Journal* (Rochester, NY), news clipping, BP.

23 W. B. Ready, "Bridey Murphy: An Irishman's View," *Magazine of Fantasy and Science Fiction* 11, no. 2 (August 1956): 81–88.

24 I have not been able to learn much about Perry Jester, but there is a photograph of him from this period in the Eugene Manis Scrapbooks at University of Florida. See http://www.uflib.ufl.edu/ufdc/?b=UF00074671&v=00001, from Eugene Manis Scrapbooks: Africa.

25 Lee Umbenhour to Bernstein, March 9, 1956, BP. Enclosed with the letter was a news clipping dated March 10, 1956.

26 Tighe to Bernstein, written on the verso of Gus Reynolds to Tighe, undated, BP, 1950s files.

27 Perry Jester to Bernstein, April 3, 1956, BP.

28 Mrs. Fred Rohr to Bernstein, January 1956, BP. In addition to Bernstein's book they read Max Freedom Long, *The Secret Science behind Miracles* (1948), *There Is a River, Many Mansions*, "all the J. B. Rhine books, Aldous Huxley, Glenn Clark"—and, she said, the "Teachings of the Far East."

29 Charles Tart to Bernstein, February 29, 1956 BP. Tart's current website is at http://www.paradigm-sys.com/.

30 Bernstein referred him to parapsychologist Ian Stevenson and kept up a correspondence with him, also beginning a correspondence with the woman Mann referred to.

31 Martin C. Mann to Bernstein, February 10, 1956 BP; Phyllis Krystal to Bernstein, January 20, 1956. Several other correspondents noted their own Cayce connections or inspirations; e.g., Grace Jones to Bernstein, March 9, 1956, BP; Albert E. Turner, of Virginia Beach, Virginia, to Bernstein, February 13, 1956, BP.

32 Doris M. Williamson to Bernstein, January 30, 1956, BP.

33 William Thomas to Bernstein, March 14, 1956, BP. Similarly, Jackie Leet to Bernstein, January 17, 1956, BP.

34 Joan Liker to Bernstein, February 3, 1956, BP.

35 "Suzette" had lived on a farm on a big river like the St. Lawrence and died trying to save her grandchild from drowning. MacIver had identified the likely home of the previous "self" on the banks of the St. Lawrence river. MacIver to Bernstein, October 24, November 20, and November 27, 1962, BP.

36 From Charles T. Frey and Louise Ireland-Frey to Bernstein, June 30, 1956, BP.

37 R. Bowerman to Bernstein, January 11, 1956, BP. I have ascertained that the University of Oregon did receive donations from this family, but nothing corresponding to this particular description. The Bowerman claims are intriguingly similar to the

lost civilization theses of 1880–1920 (Mu, Lemuria etc.) and to Mormon accounts of pre-Columbian America.

38 Irene Smith Johnson to Bernstein, January 6, 1956, BP. It seems that Johnson was seeking a kind of body memory that would have been stored in successive generations through heredity, rather than reincarnation as conventionally understood. Implicit in her question was the idea that Johnson had not lived before, but that the memories of her ancestors might have been passed down through heredity.

39 Marcus A. Williams to Bernstein, undated [1956], BP.

40 Bill Barker, "The truth about Bridey Murphy Exclusive," *Denver Post*, March 11, 1956.

41 "Bridey Murphy Puts Nation in Hypnotizzy," *Life*, March 19, 1956, 29–33. Bernstein wrote a rebuttal to this article, but *Life* refused to publish it. He also wrote an essay answering his critics, but it was never published. Copies are in the Bernstein Papers.

42 "Bridey Murphy Puts Nation in Hypnotizzy," 33.

43 See also "Strange Truths about Hypnosis, Mystery of Mind," *New York Journal American*, March 26, 1956; Watson Sims, "What Did Bridey Prove? Psychologists Discount Rebirth Idea," *New York Journal American*, March 11, 1956.

44 Bernstein to Tighe, April 18, 1956, BP.

45 Tighe to Bernstein, January 21, 1956, BP.

46 "The Strange Search for Bridey Murphy," *Denver Post*, September 12, 1954, 6–10.

47 Wally White, Wesley Hartzell, and Bob Smith, "Was Life Here Her 'Ireland'?" *Chicago American*, May 27, 1956, 1.

48 "The Bridey Murphy Controversy," *Chicago American*, May 27, 1956, 1.

49 "The Story of Bridey in Chicago," *Chicago American*, May 27, 1956, 8.

50 "Bridey's 'Uncle' Here!" *Chicago American*, May 18, 1956, 1; "Bridey's Chicago Uncle," *Chicago American*, May 28, 1956.

51 Ernest Tucker and Norman Glubok, "Bridey's Scanty 'Facts' on Ireland," *Chicago American*, June 5, 1956.

52 "Bridey Murphy's Chicago 'Brogue,'" *Chicago American*, May 29, 1956; "Bridey Was Spanked—in Chicago," *Chicago American*, May 30, 1956.

53 "More about Chicago's Bridey," *Chicago American*, May 31, 1956.

54 "'Bridey' Buried in Shroud of Facts" (last of a series), *Chicago American*, June 7, 1956.

55 Effie Alley, "Experts Hail Exposure of Bridey Myth," *Chicago American*, June 6, 1956.

56 Bernstein rebutted this in the Pocket Books edition of Bernstein's *Search for Bridey Murphy* (1956), chap. 20, "Debunking the Debunkers."

57 Alley, "Experts Hail Exposure of Bridey Myth." Second quotation is from *Chicago American*, May 27, 1956, 8.

58 Ernest Tucker/Roberta Westwood, "I Was under a Hypnotic Spell," *Chicago American*, June 3, 1956, 10.

59 Bernstein to Tighe, undated [1956]; Tighe to Bernstein, May 26, 1956, BP.

60 Bernstein to Tighe, October 22, 1956, BP, quoting reviews from all the national and

many regional papers; Bernstein to Tighe, July 23, 1956, BP, mentioning "a synopsis of the . . . movie, *You Have Lived Before.*"

61 Bernstein to Tighe, February 5, 1957, BP.

62 Tighe to Bernstein, September 19, 1961, BP.

63 Tighe to Bernstein, February 16, 1966, BP, and Tighe to Bernstein, received March 12, 1966, BP. The tone of these letters suggests that Tighe's invitation to the show was a favor to Bernstein, not something the show's directors really wanted; Bernstein helped finance her trip, as did Doubleday.

Notes to Chapter Six

1 John M. Macdonald, *Psychiatry and the Criminal: A Guide to Psychiatric Examinations for the Criminal Courts* (Springfield, IL: Thomas, 1958), 197–98.

2 *Cornell v. Superior Court of San Diego* 52 Ca.2d 99 (May 1959) [L.A. No. 25328. In Bank, May 5, 1959]. Justice Peters asked why hypnosis should "be held away as a means of compiling a case." The district attorney replied that he had "nothing against hypnosis"; it was "just a little unusual." Peters was unsatisfied with the conflation of "unusual" and "unlawful." This chapter is much indebted here and throughout to two sources on hypnosis and the law, Jean-Roche Laurence and Campbell Perry, *Hypnosis, Will, and Memory: A Psycho-legal History* (New York: Guilford Press, 1988), and Alan Scheflin and Jerrold Lee Shapiro, *Trance on Trial* (New York: Guilford Press, 1989).

3 Marvin Belli, *Blood Money: Ready for the Plaintiff* (first published in 1956) and *Modern Trials* (1954–60).

4 F. Lee Bailey, *The Defense Never Rests* (New York: Signet, 1972).

5 William J. Bryan, *Legal Aspects of Hypnosis* (Springfield, IL: Thomas, 1962).

6 See Susan L. Carruthers, "*The Manchurian Candidate* (1962), and the Cold War Brainwashing Scare," *Historical Journal of Film, Radio, and Television* 18 (1998): 75.

7 R. Cort Kirkwood, "Manchurian Candidates: North Koreans and Soviets Experimented on U.S. POWs," *Insight on the News* 12 (October 21, 1996): 7; "Action— Manchurian Myth Debunked! Sinatra, JFK, and an Age-Old Hollywood Story," *Premiere*, July 2004, 39.

8 Eugene Kinkead, *In Every War but One* (New York: Norton, 1959).

9 Edward Hunter, *Brainwashing* (New York: Pyramid Books, 1956).

10 Harvey Weinstein, *Father, Son and CIA* (Halifax, NS: Goodread Biographies, 1990), chap. 9.

11 Ibid.

12 One reviewer found the story not "plausible." *Films and Filming* 3 (1962): 37.

13 Also M. Brenman, "Experiments in the Hypnotic Production of Anti-social and Self-Injurious Behavior," *Psychiatry* 5 (1942): 49–61, and W. R. Wells, "Experiments in the Hypnotic Production of Crime," *Journal of Psychology* 11 (1941): 63–102. Carruthers, "*Manchurian Candidate*," quoted a review describing the film as a "reel-life nightmare, which could easily become a real-life one, judging by the . . . experiments in brainwashing and hypnosis one hears about."

14 Hunter, *Brainwashing*; also Edward Hunter, *Brainwashing in Red China: The Calculated Destruction of Men's Minds* (New York: Vanguard, 1951). John Marks has

argued that the techniques depicted in the film were closer to "CIA-sponsored research into 'mind control' techniques than to anything experienced by western POWs in Korea." John Marks, *The Search for the "Manchurian Candidate": The CIA and Mind Control; The Secret History of the Behavioral Sciences* (New York: Times Books, 1979).

15 Kinkead, *In Every War* but *One*. See also David Seed, *Brainwashing: The Fictions of Mind Control; A Study of Novels and Films* (Kent, OH: Kent State University Press, 2004); and Louis Menand, "Brainwashed: Where *The Manchurian Candidate* Came From," *New Yorker*, September 15, 2003.

16 Carruthers, "*Manchurian Candidate.*"

17 There have been speculations that the film played a role in the Kennedy assassination. E.g., John Loken, *Oswald's Trigger Films: "The Manchurian Candidate," We Were Strangers, Suddenly?* (Ann Arbor, MI: Falcon Books, 2000), 8–9.

18 E.g., "Mission Mind Control," *ABC News*, July 10, 1979 (part 1 of 3); "Sirhan, *The Manchurian Candidate*, and CIA Mind Control Experiments" and "Evidence of Revision: The Assassination of America" (series produced by Conspiratus Ubiquitus, Etymon Productions, Sisyphus Press, 2006).

19 "Hypnosis in National Defense: Needed, a New Manhattan Project," *Journal of the American Institute of Hypnosis* (hereafter *JAIH*) 3, no. 4 (October 1962): 4–5.

20 "Roosevelt" attributed these claims to Leonard Carmichael, an eminent biopsychologist and child psychologist. Reprinted in *JAIH* 3, no. 4 (1962): 48. Edith Kermit Roosevelt, the second wife of Teddy Roosevelt, died in 1948. See "Ever Been Fearful of Mind Control? By Edith Kermit Roosevelt," reprinted from *Long Island Daily Press*, *JAIH*, June 30, 1962.

21 *JAIH* 3 (1962): 1–5. The pamphlet's authorship is a matter of continuing controversy. It is commonly attributed to L. Ron Hubbard and his associates in the Church of Scientology. See Bent Corydon, *L. Ron Hubbard: Messiah or Madman?* (Fort Lee, NJ: Barricade Books, 1996); Russell Miller, *Bare-Faced Messiah: The True Story of L. Ron Hubbard* (New York: Henry Holt, 1987); Kevin Victor Anderson, *Report of the Board of Enquiry into Scientology* (State of Victoria, Australia, 1965 and 2005); CESNUR (Centro Studi sulle Nuove Religioni), "L. Ron Hubbard, Kenneth Goff, and the 'Brain-Washing Manual' of 1955," www.cesnur.org.

22 "Busch Defense to Try Hypnosis during Trial," *Los Angeles Times*, December 5, 1960, C11: "Matthews planned to "seek a jury whose members have seen the film mentioned by Busch."

23 The irresistible impulse defense was debated in the 1950s as the model penal code was developed. See Deborah Denno, "Crime and Consciousness: Science and Involuntary Acts," *Minnesota Law Review* 87 (December 2002): 269.

24 "Busch Triple Murder Trial Set Monday," *Los Angeles Times*, December 9, 1960, B11. Matthews "said he will seek to prove to the jury that Busch, who had seen the film *Psycho* shortly before one of the murders, was innocent of criminal intent because he was still under the influence of the movie horror story." But another story was headlined "Confessed Slayer Called Cautious and Cunning" (*Los Angeles Times*, December 13, 1960, B7). He carried handcuffs, the point being that he "anticipated" his killings. See also "Busch Trial Psychologist Will Proceed," *Los Angeles Times*, December 26, 1960, B26.

25 *People v. Busch*, 56 Ca.2d 868, 16 Cal Rptr 898; 366 P.2d 314 (1961); Bryan, *Legal Aspects of Hypnosis*, 75–76; Scheflin and Shapiro, *Trance on Trial*, 59.

26 H. E. F. Donohue, "Hitchcock: Remembrance of Murders Past," *New York Times*, December 14, 1969, D23.

27 George William Rae, *Confessions of the Boston Strangler* (New York: Pyramid Books, 1967), 142–43: The examination took place at Bridgewater in the spring of 1965. Susan Kelly, *The Boston Stranglers* (New York: Birch Lane, 1995), 173–74: Witnesses "wondered how much DeSalvo had been led . . . by Dr Bryan, so forceful and domineering."

28 Gerold Frank, *The Boston Strangler* (New York: New American Library, 1966), 254ff.

29 Kelly, *Boston Stranglers*, 71, 173–74.

30 Frank, *Boston Strangler*, 255ff.

31 *State v. Nebb*, no. 39540 Ohio CP, Franklin County (May 28, 1962). See also R. T. Creager, "The Admissibility of Testimony Influenced by Hypnosis," *Virginia Law Review* 67, no. 6 (September 1981): 1203–33. The case was discussed in Bryan, *Legal Aspects of Hypnosis*, 265, and in Harry Arons, *Hypnosis in Criminal Investigation* (Springfield, IL: Thomas, 1967), 106–9. Nebb was convicted of the lesser crime of manslaughter and became eligible for parole in eight months. Robin Waterfield, *Hidden Depths: The Story of Hypnosis* (New York: Brunner-Routledge, 2003), 265; Scheflin and Shapiro, *Trance on Trial*, 52; Laurence and Perry, *Hypnosis, Will, and Memory*.

32 A case of 1967, in which a woman who killed her husband produced a mitigating contextual narrative under hypnosis, is related in Robin Wakefield, *Hidden Depths: The Story of Hypnosis* (New York: Routledge, 2003), 265. This was part of a general push to use hypnosis in evaluating intent. "Editorial: Investigating Criminal Intent," *JAIH* 2 no. 4 (October 1961): 2–3. See also *People v. Brown*, 49 Ca.2d 577, 320 P.2d 5 (1958); *People v. Jones*, 42 Ca.2d 219, 266 P.2d 38 (1954); *People v. Cartier*, 51 Ca.2d 590, 355 P.2d 114 (1959); *People v. Modesto*, 59 Ca.2d 722, 31 Cal Rptr 225 (1963).

33 E.g., Melvin Powers, *Practical Guide to Self-Hypnosis* (New York: Melvin Powers, 1961); Leslie Lecron, *Techniques of Hypnotherapy* (New York: Julian Press, 1962); Lewis R. Wolberg, *Medical Hypnosis* (New York: Grune and Stratton, 1949).

34 *JAIH* 4, no. 3 (June 1963): 41–42; *JAIH* 5, no. 4 (October 1964): 44.

35 Arons, *Hypnosis in Criminal Investigation*, 36–37.

36 Bryan, *Legal Aspects of Hypnosis*, 138–39; Hans Sutermeister, "Screen Hypnosis," *British Journal of Medical Hypnosis* 7, no. 1 (1955): 47.

37 It was called the D-13 Test. "Must pass a test to see film," reprinted from *Boston Herald*, November 9, 1963 (in *JAIH* 5, no. 1 January 1964): 46; "Hypnosis in the News," *JAIH* 2, no. 3 (July 1961): 48.

38 William J. Bryan, "The Walking Zombie Syndrome," *JAIH* 2, no. 3 (July 1961): 10–18. Bryan gave the example of soldiers who have had traumatic and sometimes near-death experiences.

39 "The Dangers of the Waking State," *JAIH* 5, no. 2 (April 1964): 38–39.

40 "Hypnosis and the Law," *JAIH* 4, no. 2 (April 1963): 48, reporting on an article in the *Los Angeles Daily Journal*, January 3, 1963.

41 Arons, *Hypnosis in Criminal Investigation*, 27 and 28.

42 Ibid., 120–24. "Hypnotist May Help Find Body: Professional Offers to Place Confessed Killer in Trance," *Los Angeles Times*, August 21, 1963, described plans for a professional hypnotist to help Hugh Macleod Pheaster, "convicted robber-abortionist," locate the body of a "young mother" he threw into a ravine in the San Bernardino Mountains.

43 Arons, *Hypnosis in Criminal Investigation*, xv.

44 Ibid., xviii.

45 "Hypnosis in the News," *JAIH* 4, no. 2 (April 1963): 45.

46 In 1961, for instance, the superintendent of the State Board of Criminal Investigation in Olympia, Washington, recently returned from a Bryan training program, sent a letter to say that several police groups in his area wanted to develop special expertise in handling witnesses. Letters to the editor, *JAIH* 2, no. 3 (July 1961): 6; letter from Denton Johnson, superintendent, State Bureau of Criminal Investigation, Olympia, Washington, April 4, 1961.

47 "The Big Move," *JAIH* 5, no. 4 (October 1964): 3–4.

48 The institute was operational for at least a quarter of a century, through at least the early 1980s. Arons's textbook was *Hypnosis in Criminal Investigation* (1967).

49 Ibid., 21 and 87. Arons reprinted the text of an article by Earl H. Cramer from the April 1964 issue of *Aerospace Medicine*, relating to an event in 1962 when a commercial airliner went down in the Canton Island area of the Pacific. Under hypnosis, the pilot was able to reconstruct events he had earlier been unable to remember, clearing himself of possible charges of negligence. I have been unable to locate the original source Arons was quoting. Arons, *Hypnosis in Criminal Investigation*, 88–93. On his training courses, see 22 and 31.

50 Arons claimed that the first "officially approved" course was in Ridgefield, NJ; Arons, *Hypnosis in Criminal Investigation*, 30.

51 George Barmann, "Solving Crimes by Hypnosis," *Popular Mechanics* 113, no. 4 (April 1960), 106–9, 234, 236.

52 *Harding v. State*, 5 Md. App. 230, 246 A.2d 302 (1968).

53 The leading figure in this period was Martin Reiser, a clinical psychologist who ran a hypnosis training institute and forensics service associated with the Los Angeles Police Department during the 1970s and early 1980s. His unpublished training materials evolved into a textbook: Martin Reiser, *Handbook of Investigative Hypnosis* (Los Angeles: Lehi, 1980).

54 Gail Miller and Sandra Tompkins, *Kidnapped! At Chowchilla* (Plainfield, NJ: Logos International, 1977); Jack W. Baugh and Jefferson Morgan, *Why Have They Taken Our Children? Chowchilla, July 15, 1976* (New York: Delacorte Press, 1978).

55 Curt R. Bartol and Anne M. Bartol, *Introduction to Forensic Psychology: Research and Application* (Thousand Oaks, CA: Sage, 2008) 37; http://www.time.com/time/magazine/article/0,9171,914589,00.html#ixzz0riIYXoiW.

56 Scheflin and Shapiro, *Trance on Trial*, 73, cite *U.S. v. Adams* 581 F. 2d 193, 198–99 (9th Cir.), cert. denied, 439 U.S. 1006, 99 S. Ct. 621, 58 L. Ed. 2d 683 (1978) as the first case to question the open admissibility policy.

57 T. J. Kaptchuk, "Intentional Ignorance: A History of Blind Assessment and Placebo Controls in Medicine," *Bulletin of the History of Medicine* 72 (1998): 389–433.

Incidents of this kind are sometimes recounted as landmark moments in the history of the placebo and of controlled experiments. On the controlled trial, see Harry M. Marks, *The Progress of Experiment: Science and Therapeutic Reform in the United States, 1900–1990* (New York: Cambridge University Press, 1997), and Kurt Danziger, *Constructing the Subject: Historical Origins of Psychological Research* (New York: Cambridge University Press, 1994).

58 E.g., R. W. White, "Preface to the Theory of Hypnosis," in *Contemporary Psychopathology: A Source Book*, ed. Silvan S. Tomkins (Cambridge, MA: Harvard University Press, 1943); P. C. Young, "Hypnotic Regression—Fact or Artifact?" *Journal of Abnormal Social Psychology* 35 (1940): 273–78; J. M. Stalnaker and E. E. Riddle, "The Effect of Hypnosis on Long-Delayed Recall," *Journal of General Psychology* 6 (1932): 429–40.

59 Martin Orne, "On the Social Psychology of the Psychological Experiment: With Particular Reference to Demand Characteristics and Their Implications," *American Psychologist* 17 (1962): 777.

60 Martin Orne, "The Nature of Hypnosis: Artifact and Essence," *Journal of Abnormal and Social Psychology* 58 (1959): 277–99.

61 E.g., Richard I. Thackray and M. T. Orne, "Effects of the Type of Stimulus Employed and the Level of Subject Awareness on the Detection of Deception," *Journal of Applied Psychology* 52, no. 3 (June 1968): 234–39.

62 E.g., S. Sherman, "Demand Characteristics in an Experiment on Attitude Change," *Sociometry* 30, no. 3 (September 1967): 246–61; P. Wuebben, "Experimental Design, Measurement, and Human Subjects: A Neglected Problem of Control," *Sociometry* 31, no. 1 (March 1968): 89–101; J. D. Martin, "Suspicion and the Experimental Confederate: A Study of Role and Credibility," *Sociometry* 33, no. 2 (June 1970): 178–92; David Finkelman, "Science and Psychology," *American Journal of Psychology* 91, no. 2 (June 1978): 179–99; J. G. Adair, "Demand Characteristics or Conformity? Suspiciousness of Deception and Experimenter Bias in Conformity Research," *Canadian Journal of Behavioural Science* 4, no. 3 (July 1972): 238–48; W. DeJong, "Another Look at Banuazizi and Movahedi's Analysis of the Stanford Prison Experiment," *American Psychologist* 30, no. 10 (October 1975): 1013–15.

63 E.g., R. G. Boutilier, J. C. Roed, and A. C. Svendsen, "Crises in the Two Social Psychologies: A Critical Comparison," *Social Psychology Quarterly* 43, no. 1 (March 1980): 5–17.

64 Michael Barnes, "The Mind of a Murderer," BBC Horizon for *Frontline*, March 19, 1984.

65 M. T. Orne, Affadatit of Amicus Curiae, *Quaglino v. People*, U.S. Sup. Ct. No. 77–1288, cert. den. 11/27/78. Also published in E. Margolin, chm., *Sixteenth Annual Defending Criminal Cases: The Rapidly Changing Practice of Criminal Law*, vol. 2 (New York: Practicing Law Institute, 1978), 831–57.

66 For a discussion of this case, and Orne's brief, in the broader context of other cases at this time, see Scheflin and Shapiro, *Trance on Trial*, 77.

67 Martin Orne, "The Use and Misuse of Hypnosis in Court," *International Journal of Clinical and Experimental Hypnosis* 27 (1979): 311–41; Scheflin and Shapiro, *Trance on Trial*, 74. Scheflin and Shapiro date judicial nervousness about hypnosis to *People v. Adams*; both Scheflin and Shapiro, *Trance on Trial*, and Lawrence and

Perry, *Hypnosis, Will, and Memory*, discuss a stream of problematic cases involving hypnosis in the late 1970s that culminated in the Shirley case.

68 *State v. Hurd*, N.J. 86 N.J. 525 (1981), discussed in Scheflin and Shapiro, *Trance on Trial*, 78; this ruling has been overturned recently, in *State v. Moore*, 188 N.J. 182, 902 A.2d 1212 (2006). Thanks to Alan Scheflin for this reference.

69 Scheflin and Shapiro, *Trance on Trial*, 91.

70 *People v. Shirley*, 31 Cal. 3d 18.

71 Ibid., 90.

72 Diamond, Cal. Law Rev 68:2 March 1980, 313–49. In the years 1980–82, several states made a per se rule against hypnosis—Shirley was only the most influential, because it was in California. For a detailed account see Scheflin and Shapiro, *Trance on Trial*, 75–95.

73 Scheflin and Shapiro, *Trance on Trial*, 80; Reiser, *Handbook of Investigative Hypnosis* (Los Angeles: Lehi, 1980).

Notes to Chapter Seven

1 E.g., Barbara Kirshenblatt-Gimblett, "Kodak Moments, Flashbulb Memories: Reflections on 9/11," *Drama Review* 47, no. 1 (2003): 11–48; Dorthe Kirkegaard Thomsen and Dorthe Berntsen, "Snapshots from Therapy: Exploring Operationalisations and Ways of Studying Flashbulb Memories for Private Events," *Memory* 11, no. 6 (2003): 559–70.

2 "Where Were You?" *Esquire*, November 1973, 136–37.

3 Roger Brown and James Kulik, "Flashbulb Memories," *Cognition* 5 (1977): 73–99.

4 Mark Feeney, "40 Years after Those 6.9 Seconds in Dallas," *Boston Globe*, November 22, 2003.

5 George Lipsitz, *Time Passages: Collective Memory and American Popular Culture* (Minneapolis: University of Minnesota Press, 2001).

6 Flora Rheta Schreiber, "Television's New Idiom in Public Affairs," *Hollywood Quarterly* 5, no. 2 (1950): 144–52; Irving Kahn, "TV: The Second Decade," *Analysts Journal* 14, no. 1 (1958): 67–70; Dallas W. Smythe, "The Consumer's Stake in Radio and Television," *Quarterly of Film Radio and Television* 6, no. 2 (1951): 109–28.

7 Archival Television Audio, www.atvaudio.com/jfk.php.

8 It seems likely from the survey questions that they considered and rejected adding a Latino informant group to their study, but I have not been able to confirm this.

9 See David Wrone, *The Zapruder Film: Reframing JFK's Assassination* (Lawrence: University Press of Kansas, 2003), and more recently Max Holland and Johann Rush, "J.F.K.'s Death, Re-framed," *New York Times*, November 22, 2007.

10 There was a certain irony to Cartier-Bresson's "moment," because he used a Leica, which took a stream of photographs in quick succession, producing a whole series of "moments." Photojournalists remember Cartier-Bresson's using the Leica to produce natural, spontaneous images rather than the staged shots common before his time; see *Henri Cartier-Bresson: The Man, the Image and the World; A Retrospective*, ed. Philippe Arbaïzar et al. (London: Thames and Hudson, 2003).

11 In his 1955 book *The Europeans* (New York: Simon and Schuster).

12 John Wolbarst, *Pictures in a Minute: How to Get the Most Out of Your Polaroid Land*

Camera (New York: American Photographic Book Publishing, 1956). For the linguistic shift see Wolbarst's monthly columns in *Modern Photography*, 1959–66.

13 The advertisement played on the idea of spontaneity: http://www.youtube.com/watch?v=h7k2uwJmwxo.

14 The Polaroid One Step, market leader for years, was launched in 1977; see Nicole Columbus, ed., *Innovation/Imagination: 50 Years of Polaroid History* (New York: Abrams, 1999), 119.

15 Peter Schjedahl, "The Instant Age," in *Legacy of Light*, ed. Constance Sullivan (New York: Knopf, 1987), 8–13; *Life*, October 27, 1970.

16 The first image in Roland Barthes's *Camera Lucida* was a Polaroid, a choice that has inspired much critical discussion, although Barthes himself found the medium disappointing. For Heath's discussion see "Rhetoric of the Image," in Roland Barthes and Stephen Heath, *Image—Music—Text*, ed. and trans. Stephen Heath (New York: Hill and Wang, 1976). On the definitive 1972 camera, see Geoffrey Batchen, ed., *Photography Degree Zero: Reflections on Roland Barthes's "Camera Lucida"* (Cambridge, MA: MIT Press, 2009), 34, 38.

17 Robert B. Livingston, "Reinforcement" and "Brain Circuitry related to Complex Behavior," in *The Neurosciences: A Study Program*, ed. G. Quarton, T. Melnechuk, and F. Schmitt (New York: Rockefeller Press, 1967), 568–76, 499–515.

18 "Brain Mechanisms in Conditioning and Learning: A Report of an NRP Work Session," *Neurosciences Research Program Bulletin* 4, no. 3 (1966): 235–354; "Arousal Effects on Evoked Activity in a 'Nonsensory' System," *Science* 139, no. 3554 (1963): 502–4.

19 Livingston, "Reinforcement," 576.

20 Livingston, "Brain Circuitry," 512–15.

21 Allan N. Schore, *Affect Regulation and the Origin of the Self: The Neurobiology of Emotional Development* (Hillsdale, NJ: Erlbaum, 1999).

22 Eric Nagourney, "Neal E. Miller Is Dead at 92; Studied Brain and Behavior," *New York Times*, April 2, 2002, A21.

23 Neal E. Miller, "Some Reflections on the Law of Effect Produce a New Alternative to Drive Reduction," in *Nebraska Symposium on Motivation, 1963*, ed. Marshall R. Jones (Lincoln: University of Nebraska Press, 1963), 65–112.

24 Livingston, "Brain Mechanisms in Conditioning and Learning," *Neurosciences Research Progress Bulletin* 4, no. 3 (1985): 258; Miller, "Some Reflections on the Law of Effect." See also Stephen Grossberg, *The Adaptive Brain: Cognition, Learning, Reinforcement, and Rhythm*, 2 vols. (Amsterdam: North-Holland, 1987), 2:27.

25 Robert B. Livingston, statement (November 18, 1993), in *Assassination Science: Experts Speak Out on the Death of JFK*, ed. James H. Fetzer (Chicago: Open Court Press, 1998), 162–71. All quotations and references relating to Livingston's thoughts on the Kennedy assassination are drawn from this paper.

26 Brown and Kulik, "Flashbulb Memories," 98.

27 Eugene Winograd and Ulric Neisser, "Affect and Accuracy in Recall: Studies of 'Flashbulb' Memories," *Imagination, Cognition, and Personality* 13, no. 4 (1994): 367; Martin A. Conway, *Flashbulb Memories* (Hove, UK: Erlbaum, 1995); Anne Turyn and Andy Grundberg, *Missives* (New York: Alfred Van der Marck, 1986); "Flashbulb Memories: The Picture Fades," *Science News* 133, no. 23 (June 4, 1988): 358.

28 David B. Pillemer, "Flashbulb Memories of the Assassination Attempt on President Reagan," *Cognition* 16, no. 1 (February 1984): 63–80.

29 Jennifer M. Talarico and David C. Rubin, "Confidence, Not Consistency, Characterizes Flashbulb Memories," *Psychological Science* 14, no. 5 (2003): 455–61; J. M. Talarico and D. C. Rubin, "Flashbulb Memories Are Special After All: In Phenomenology, Not Accuracy," *Applied Cognitive Psychology* 21, no. 5 (2007): 557–78; E. Krackow, S. J. Lynn, and D. G. Payne, "The Death of Princess Diana: The Effects of Memory Enhancement Procedures on Flashbulb Memories," *Imagination, Cognition, and Personality* 25, no. 3 (2006): 197–220; "Flashbulb Memories: Confident Blunders," *Science News* 143 (March 13, 1993): 11.

30 The baseball game memory would itself become the subject of controversy, as psychologists weighed in on just how good or bad one ought to consider the memory to have been.

31 Eugene Winograd and Ulric Neisser, eds., *Affect and Accuracy in Recall: Studies of "Flashbulb" Memories* (Cambridge: Cambridge University Press, 1992).

32 Ulric Neisser, ed., *Memory Observed: Remembering in Natural Contexts* (San Francisco: Freeman, 1982).

33 J. N. Bohannon, "Flashbulb Memories for the Space Shuttle Disaster: A Tale of Two Theories," *Cognition* 29, no. 2 (1988): 179–96; Michael McCloskey, Cynthia G. Wible, and Neal J. Cohen, "Is There a Special Flashbulb-Memory Mechanism?" *Journal of Experimental Psychology: General* 117, no. 2 (1988): 171–81.

34 Ulric Neisser and Nicole Harsch, "Phantom Flashbulbs," in *Affect and Accuracy in Recall: Studies of "Flashbulb" Memories* (Atlanta: Emory University, 1992), 9–31.

35 David M. Diamond, Adam M. Campbell, Collin R. Park, Joshua Halonen, and Phillip R. Zoladz, "The Temporal Dynamics Model of Emotional Memory Processing: A Synthesis on the Neurobiological Basis of Stress-Induced Amnesia, Flashbulb and Traumatic Memories, and the Yerkes-Dodson Law," *Neural Plasticity* 2007 (2007): 60803.

36 James L. McGaugh, "Time-Dependent Processes in Memory Storage," *Science* 153 (1966): 1351–58; J. L. McGaugh and M. J. Herz, *Memory Consolidation* (San Francisco: Albion, 1972); James L. McGaugh, "Memory: A Century of Consolidation," *Science* 287 (2000): 248–51.

37 C. Finkenauer, O. Luminet, L. Gisle, A. El-Ahmadi, M. Van Der Linden, and P. Philippot, "Flashbulb Memories and the Underlying Mechanisms of Their Formation: Toward an Emotional-Integrative Model," *Memory and Cognition* 26 (1998): 516–31; Cheryl M. Paradis, Faith Florer, Linda Zener Solomon, and Theresa Thompson, "Flashbulb Memories of Personal Events of 9/11 and the Day After," *Psychological Reports* 95, no. 1 (2004): 304; I. Nachson and A. Zelig, "Flashbulb and Factual Memories: The Case of Rabin's Assassination," *Applied Cognitive Psychology* 17, no. 5 (2003): 519–32; Martin A. Conway, *Flashbulb Memories* (Hillsdale, NJ: Erlbaum, 1995).

38 Lenore Terr, "Children of Chowchilla," *Psychoanalytic Study of the Child* 34 (1979): 547–623; Lenore C. Terr, "Chowchilla Revisited: The Effects of Psychic Trauma Four Years after a School-Bus Kidnapping," *American Journal of Psychiatry* 140 (1983): 1543–50.

39 Lenore Terr, "Children's Memories in the Wake of *Challenger*," *American Journal*

of Psychiatry 153 (1996): 618–25; Lenore Terr, "Childhood Traumas: An Outline and Overview," *American Journal of Psychiatry* 148 (1991): 13.

40 Daniel L. Schachter, *Searching for Memory: The Brain, the Mind, and the Past* (New York: Basic Books, 1996), 256.

41 E.g., Mauricio Sierra and German E. Berrios, "Flashbulb Memories and Other Repetitive Images: A Psychiatric Perspective," *Comprehensive Psychiatry* 40, no. 2 (1999): 115–25.

42 Ruth Leys, *Trauma: A Genealogy* (Chicago: University of Chicago Press, 2000), 259, quoting Van der Kolk and then commenting on his text. For her critical examination of such projects more generally see chap. 9, "The Science of the Literal"; Bessel A. Van der Kolk, Alexander C. McFarlane, and Lars Weisaeth, *Traumatic Stress: The Effects of Overwhelming Experience on Mind, Body, and Society* (New York: Guilford Press, 1996), 297; Cathy Caruth, *Unclaimed Experience: Trauma, Narrative, and History* (Baltimore: Johns Hopkins University Press, 1996), and Cathy Caruth, *Trauma: Explorations in Memory* (Baltimore: Johns Hopkins University Press, 1995), 152–53. For other critical discussion of the idea of traumatic memories as literal truth, see Ian Hacking, *Rewriting the Soul: Multiple Personality and the Sciences of Memory* (Princeton, NJ: Princeton University Press, 1995), 253–54.

Notes to Chapter Eight

1 Carey Mulligan, "The Monster in the Mists," *New York Times*, May 15, 1994.

2 Sandra Butler, *Conspiracy of Silence: The Trauma of Incest* (Volcano, CA: Volcano Press, 1996); Louise Armstrong, *Kiss Daddy Goodnight* (New York: Hawthorn Press, 1978); Judith Herman, *Father-Daughter Incest* (Cambridge, MA: Harvard University Press, 1982).

3 Alice Miller, *Banished Knowledge: Facing Childhood Injuries*, 3rd ed. (New York: Anchor Press, 1988), 184.

4 Indeed, this statement itself became part of Lorey Newlander's cause of action: she added slander to her list of complaints (the defense counsel argued that this claim should be rejected on grounds that it was true and so not slanderous, but the court did not accept this).

5 Myrna Oliver, "ACLU Will Pursue Appeal in Legal Test of Incest Issue," *Los Angeles Times*, August 10, 1983, C6.

6 *Newlander v. Newlander*, No. C 31815 (Cal. Super. Ct., Los Angeles County, May 26, 1983 (sustaining demurrer to the complaint based on statute of limitations), appeal docketed, No. B003156 (Cal. Ct. App., 2d App. Dist., January 9, 1984).

7 Oliver, "ACLU Will Pursue Appeal."

8 Telephone interview with Melissa Salten, December 21, 2008.

9 Melissa G. Salten, "Statutes of Limitations in Civil Incest Suits: Preserving the Victim's Remedy," *Harvard Women's Law Journal* 7 (1984): 189; Angelo Rosa, "Litigating Adult Claims of Childhood Sexual Abuse," *Los Angeles Lawyer* 30, no. 6 (September 2007): 12–18; "Adult Incest Survivors and the Statute of Limitations: The Delayed Discovery Rule and Long-Term Damages," *Santa Clara Law Review* 25 (Winter 1985): 191–225.

10 Stephen Nohlgren, "Making a Case to Punish Incest," *Tampa Bay and State*, April 28, 1991, 1B.

11 Bill Corcoran, telephone interview, December 2008.

12 The movie aired in 1993 and has been regularly shown since then on cable networks.

13 Paul Eberle, *The Abuse of Innocence: The McMartin Preschool Trial* (Amherst, NY: Prometheus Books, 1993).

14 Quoted in Rebecca Thomas, "Adult Survivors of Childhood Sexual Abuse and Statutes of Limitations: A Call for Legislative Action," *Wake Forest Law Review* 26 (1991): 1248 n. 301. The bill was AB 2323. Elihu Mason Harris is a former United States Democratic party politician. He served for twelve years (1978–91) as a member of the California State Assembly before his election as Oakland mayor.

15 I have not been able to discern the specific differences between the original bill and the final legislation.

16 *DeRose v. Carswell*, 196 Cal.App.3d 1011, 242 Cal.Rptr. 368, 56 USLW 2452 (review denied March 17, 1988).

17 *Mary D. v. John D.* 216 Cal. App. 3d 285, 264 Cal. Rptr. 633 (1989).

18 264 Cal.Rptr. at 636 (quoting Cal.Civ.Proc.Code § 340.1(d) (West 1990)).

19 *Petersen v. Bruen*, 106 Nev. 271, 792 P.2d 18, 20 (1990) (citing 51 Am.Jur.2D Limitation of Actions § 19 (1970), which cites *Chase Sec. Corp. v. Donaldson*, 325 U.S. 304, rehearing denied, 325 U.S. 896 (1945)).

20 The article cited by the majority was Wesson, "Historical Truth, Narrative Truth, and Expert Testimony," 60 Wash.L.Rev. 331 (1985). Id. at 86, 727 P.2d at 233.

21 The article also made interesting use of the Patty Hearst case: "Testimony of Dr. Martin Orne," in *The Trial Of Patty Hearst*, ed. Kenneth J. Reeves (San Francisco: Great Fidelity Press, 1976), 294–300.

22 Barbara Laker, "A Nightmare of Memories: Yesterday's Secrets Stir Up Present-Day Horrors for Families," *Seattle Post-Intelligencer*, April 14, 1992, C1.

23 Wash. Rev. Code § 4.16.350 (1988).

24 As long as they used pseudonyms to protect defendants' privacy and provided affidavits from attorneys and mental health professionals stating that their claims were not frivolous.

25 "Remembering," False Memory Syndrome Foundation Archives (Center for Inquiry, Amherst, NY), box 17, survivor newsletters.

26 Quotation is from Connie Kristiansen, a psychologist at Carleton University. Thunder Bay newspaper clipping, False Memory Syndrome Foundation Archives, box 17, survivor newsletters.

27 Clipping, False Memory Syndrome Foundation Archives, box 17, survivor newsletters.

28 Abused children appeared to grow up, but hidden away was an "inner child." This concept came from John Bradshaw (e.g., Bradshaw, *Homecoming: Reclaiming and Championing Your Inner Child* [New York: Bantam Books, 1990]). Survivors of Incest Anonymous (SIA) explained that its members often talked about "getting in touch with their little girl." SIA, "Honoring the Child Within" (1986), False Memory Syndrome Foundation Archives, box 17.

29 *Bradshaw on: The Family: A New Way of Creating Solid Self-Esteem* (PBS Television,

1985); *Bradshaw on: Homecoming: Reclaiming and Championing Your Inner Child* (PBS Television, 1990); and *Healing the Shame That Binds You* (PBS Television, 1987).

30 SIA, "Honoring the Child Within" (1986), FMSF Archives, box 17.

31 "Satanic Ritual Abuse: A Survivor's Story," signed "Kim and Kimmie" (1992), FMSF Archives, box 17.

32 *The Scream*, call for submissions from Lori Antonitto, Sean Madden, and Eileen D'Esterno, FMSF Archives, box 17.

33 *Surviving Spirit*, FMSF Archives, box 17.

34 Survivors Oppose Ubiquitous Perps, October 1993, 34. SOUP asked its readers to write to a local paper and to a district attorney about Ralph Underwager (on Underwager, see chapter 10), FMSF Archives, box 17.

35 "The Story behind Write to Heal," *Write to Heal*, October 1993, FMSF Archives, box 17.

36 *Write to Heal*, November 1993–January 1994, 3–4. Anne M. Cox, the journal's editor, described, rather resentfully, an exchange with a therapist in which the therapist encourages her to move on, to stop calling herself a survivor and start calling herself a "thriver." Ambivalence about therapists and therapy was a significant theme of this journal at this time.

37 "Courts Begin to Respect Memory of Child Abuse," *New York Times*, January 8, 1991, A17.

38 Thomas J. Lueck, "Sharing Horrors of Childhood Sexual Abuse, 3 Join Legal Debate," *New York Times*, May 5, 1992, B1.

39 Harry MacLean, *Once Upon a Time: A True Story of Memory, Murder, and the Law* (New York: HarperCollins, 1993). According to MacLean (p. 106, but not supported by references), Eileen Lipsker told her mother "that she had visualized the killing while under hypnosis." For opposing versions of the Franklin case compare MacLean's book with Lenore Terr, *Unchained Memories: True Stories of Traumatic Memories, Lost and Found* (New York: Basic Books, 1994).

40 For instance, see W. G. Sebald's discussion of Thomas Brown's *Urn Burial* in Sebald's *The Rings of Saturn* (New York: New Directions, 1999), 19–26.

Notes to Chapter Nine

1 Broadcast by NBC, December 19, 1974.

2 Robert Buckhout, "Nearly 2000 Witnesses Can Be Wrong," *Bulletin of Psychonomic Science* 16 (1980): 307–10.

3 Bruce Bower, "Remembrance of Things False: Scientists Incite Illusory Memories and Explore Their Implications," *Science News*, August 24, 1996.

4 He picked up a variety of miscellaneous psychological writings by Stout, James Ward's entry on psychology in the *Encyclopaedia Britannica* (1886), and C. S. Myers's *Manual of Experimental Psychology* (1909).

5 See John Forrester, "1919: Psychology and Psychoanalysis, Cambridge and London—Myers, Jones and Maccurdy," *Psychoanalysis and History* 10, no. 1 (2008): 37–94. Bartlett's brief account of his early life is in Carl Murchison, ed., *A History of Psychology in Autobiography*, vol. 3 (Worcester, MA: Clark University Press, 1936).

6 Other researchers included Ward, nearing retirement but still active, the social psychologist William McDougall, and Henry Head, a renowned neurologist based in London who collaborated with Rivers in Cambridge. See L. S. Hearnshaw, *A Short History of British Psychology, 1840–1940* (New York: Barnes and Noble, 1964), chap. 3.

7 Ian Langham, *The Building of British Social Anthropology: W. H. R. Rivers and His Cambridge Disciples in the Development of Kinship Studies, 1898–1931* (London: Reidel, 1981), 65; Anita Herle and Sandra Rouse, eds., *Cambridge and the Torres Strait: Centenary Essays on the 1898 Anthropological Expedition* (Cambridge: Cambridge University Press, 1988); on Rivers's work see Richard Slobodin, *W. H. R. Rivers: Pioneer Anthropologist, Psychiatrist of "The Ghost Road"* (Sutton, UK: Phoenix Mill, 1978), 98–113, 169–78.

8 On the moral sciences tripos at this time, see J. P. Forrester, "Changing the Humanities/The Humanities Changing," unpublished paper, Gillespie Conference Center, Cambridge, July 2009.

9 A. C. Haddon, *The Decorative Art of New Guinea: A Study in Papuan Ethnography* (London, 1894); A. C. Haddon, *Evolution in Art as Illustrated by the Life History of Designs* (London: Scott, 1914).

10 W. H. R. Rivers, *History of Melanesia*, 2 vols. (Cambridge: Cambridge University Press, 1914), 2:383.

11 According to Mary Douglas, *How Institutions Think* (Syracuse, NY: Syracuse University Press, 1986), chap. 7.

12 Frederic Bartlett, "Transformations Arising from Repeated Representation: A Contribution towards an Experimental Study of the Process of Conventionalization," fellowship diss., St. John's College, Cambridge.

13 Frederic Bartlett, *Thinking: An Experimental and Social Study* (London: Allen and Unwin, 1957), chap. 8, sec. 3; Jean Philippe, "Sur les transformations de nos images mentales," *Revue Philosophique* 41 (1897): 481–93.

14 Bartlett, *Thinking*, chap. 8, sec. 3.

15 Bartlett cited in Frank Smith, "An Experimental Investigation of Perception," *Journal of Psychology* 6 (1914): 321–62. Smith argued that subjects seemed initially to get a general impression of a picture and subsequently examined the details of this mental image analytically. On the tachistoscope see Ruth Benschop, "What Is a Tachistoscope? Historical Explorations of an Instrument," *Science in Context* 11 (1998): 23–50. This instrument was first developed to test how brief a stimulus could arouse visual sensation, but it became adapted for a variety of purposes in laboratory psychology that required stimuli at the lower limits of visual perception.

16 Bartlett discussed this research in "An Experimental Study of Some Problems of Perceiving and Imaging," *British Journal of Psychology* 8 (1916): 222–66.

17 Bartlett cited John Oscar Quantz, *Problems in the Psychology of Reading* (London: Oxford University Press, 1897).

18 Bartlett never explained how he chose the painting. It is possible that its narrative character and its realism were a factor. Yeames, whose work was no longer in fashion, died in 1918, bringing some renewed publicity to his work. Obituary, *Times* (London), May 8, 1918.

19 Bartlett, "Experimental Study of Some Problems of Perceiving and Imagining," 247–48.

20 Ibid., 249, 252–54.

21 George V. Dearborn, "A Study of Imaginations," *American Journal of Psychology* 9, no. 2 (January 1898): 183–90; Kirkpatrick, "Individual Tests of School Children," *Psychological Review* 7, no. 3 (1900): 274–80; Stella Emily Sharp, "Individual Psychology," *American Journal of Psychology* 10, no. 3 (April 1899): 329–91. Bartlett would later add Rorschach (*Psychodiagnostik* [1921]) among others when he updated his references, but he did not change his analytical framework. On the use of inkblots to study the imagination in the late 1910s see Peter Galison, "Image of Self," in *Things That Talk: Object Lessons from Art and Science*, ed. Lorraine Daston (New York: Zone, 2004), 259–63. Beginning in the 1930s or a little earlier, Rorschach's *Psychodiagnostik* would be used as a clinical technology by psychiatrists and clinical psychologists, eventually becoming very popular among clinicians.

22 Frederic Bartlett, "The Functions of Images," *British Journal of Psychology* 11 (1921): 320–37, and Bartlett, "The Relevance of Visual Imagery to the Process of Thinking: Pt. III," *British Journal of Psychology* 18 (1927): 23–29.

23 Bartlett, "Experimental Study of Some Problems of Perceiving and Imagining," 253.

24 Bartlett, "Transformations Arising from Repeated Representation," 34–35.

25 Alfred Binet, *La suggestibilité* (Paris: Doin, 1900), chaps. 2–4.

26 Bartlett, "Transformations Arising from Repeated Representation," 85ff.

27 For his first descriptions of these methods, see Bartlett, "Transformations Arising from Repeated Representation," 24–29.

28 Franz Boas, "Kathlamet Texts," *Bureau of American Ethnology Bulletin* 26 (1901): 185.

29 Frederic Bartlett, *Psychology and Primitive Culture* (Cambridge: Cambridge University Press, 1923).

30 W. H. R. Rivers, *Psychology and Ethnology* (London: Kegan Paul, 1926 [posthumous]), 23–24.

31 W. H. R. Rivers, *The Todas* (London: Macmillan, 1906), 20–21.

32 With some interesting exceptions. For instance, Bartlett also tested the serial recall of a group of Indian graduate students for an African folk story. During the first round, their responses were like those of the British subjects. But a second round of tests revealed that the Indians made more "elaborations" of the tale and differed from the British subjects in the points they emphasized.

33 See Forrester, "Changing the Humanities."

34 Bartlett claimed that his view of mind as a single complex system was also indebted to James Ward and G. F. Stout. See "Return to Remembering 1968," unpublished manuscript, Cambridge University Library Archives.

35 I have not been able to identify the hospital; it was not Addenbrooke's, the local Cambridge hospital.

36 F. C. Bartlett, "What's the Use of Psychology?" unpublished manuscript (1964), Cambridge University Library (MS ref: Add. 8076).

37 Bartlett, "Transformations Arising from Repeated Representation."

38 Bartlett, "Experimental Study of Some Problems of Perceiving and Imagining";

Bartlett, "The Functions of Images," *British Journal of Psychology* 11 (1920): 320–37; Bartlett, "Feeling, Imaging and Thinking," *British Journal of Psychology* 16 (1925): 16–28.

39 Bartlett, *Thinking*, chap. 8, sec. 3; Jaan Valsiner and Alberto Rosa, *The Cambridge Handbook of Sociocultural Psychology* (Cambridge: Cambridge University Press, 2007), 663.

40 See Forrester, "1919."

41 Frederic Bartlett, "Some Experiments on the Reproduction of Folk Stories," *Folk-Lore* 31 (1920): 30–47; "Psychology in relation to the Popular Story," *Folk-Lore* 31 (1920): 264–93; "The Functions of Images," *British Journal of Psychology* 11 (1921): 320–37; "An Experiment upon Repeated Reproduction," *Journal of General Psychology* 1 (1928): 54–63.

42 This was a quotation from Franz Ricklin, *Wish Fulfilment and Symbolism in Fairy Tales* (New York: Nervous and Mental Disease, 1967). Bartlett discussed Paul Hermant, *Concerning the Fantastic in Popular Tales*, and the most sophisticated and influential articulation of this argument, by Otto Rank, *Myth of the Birth of the Hero: A Psychological Exploration of Myth*, trans. F. Robbins and Smith Ely Jelliffe (New York: *Journal of Nervous and Mental Diseases*, 1914).

43 He later addressed the issue of symbolism directly: Frederic Bartlett, "Symbolism in Folk-Lore," in *Proceedings of the Seventeenth International Congress of Psychology* (Cambridge: Cambridge University Press, 1924), 278–89.

44 His primary example was Franz Boas, *Comparative Study of Tsimshian Mythology*, Thirty-first Annual Report *of the Bureau of American Ethnology* (Washington DC: Government Printing Office, 1916), esp. 872ff. He also cited A. M. Hocart, "The Common-Sense of Myth," *American Anthropologist* 18, no. 3 (July–September 1916): 307–18, and Alice Cunningham Fletcher, James R. Murie, and Edwin S. Tracy, *The Hako: A Pawnee Ceremony* (Washington, DC: Government Printing Office, 1904).

45 Bartlett, *Psychology and Primitive Culture*, ix, 294. See also Bartlett, "Group Organisation and Social Behaviour," *International Journal of Ethics* 35 (1925): 346–67, and Bartlett, "Psychology of Culture Contact," *Encyclopaedia Britannica* (1926), 1:765–71.

46 Lucien Lévy-Bruhl, *Primitive Mentality* (New York: Macmillan, 1923).

47 William McDougall's *Introduction to Social Psychology* (London: Macmillan, 1908); the anthropological references mainly related to a trip Bartlett took to South Africa in 1923–24.

48 See David Bloor, "Whatever Happened to Social Constructiveness?" in Frederic Bartlett, *Culture, and Cognition*, ed. Akiko Saito (Hove, UK: Psychology Press, 2000), 194–215, and Bartlett, *Remembering*, 276–77.

49 Hermann Munk, *Über die Funktionen der Grosshirnrinde* (Berlin: Hirschwald, 1881; 2nd ed., 1890).

50 Henry Head, *Studies in Neurology* (Oxford: Oxford University Press, 1920), 2:605–6.

51 Ibid.; cited in Bartlett, *Remembering* (1932), 199.

52 F. C. Bartlett, "The Relevance of Visual Imagery to the Process of Thinking, Part 3," *British Journal of Psychology* 18 (1927): 23–29, at 24–25.

53 Bartlett, "Critical Notice of Head's *Aphasia*," *Brain: A Journal of Neurology* 49 (1926): 585–86.

54 Bartlett, *Remembering*, 198.

55 Ibid., 201. For more on Bartlett's concept of schema see William Brewer, "Bartlett's Concept of the Schema and Its Impact on Theories of Knowledge Representation in Contemporary Cognitive Psychology," in Frederic Bartlett, *Culture and Cognition*, ed. Akiko Saito (Hove, UK: Psychology Press, 2000), 69–89.

56 Bartlett, *Remembering*, 201–2.

57 E.g., Elizabeth B. Johnston, "The Repeated Reproduction of Bartlett's *Remembering*," *History of Psychology* 4, no. 4 (2001): 341–66; James Ost and Alan Costall, "Misremembering Bartlett: A Study in Serial Reproduction," *British Journal of Psychology* 93, no. 2 (2002): 243–55; D. E. Beals, "Reappropriating Schema: Conceptions of Development from Bartlett and Bakhtin," *Mind, Culture, and Activity* 5 (1998): 3–24.

58 Ulric Neisser and Eugene Winograd, eds., *Remembering Reconsidered: Ecological and Traditional Approaches to the Study of Memory* (Cambridge: Cambridge University Press, 1995), 2.

59 For an argument that Bartlett is misrepresented by claims about faulty memories, not for the reason I give (that he was not interested in the dichotomy between truth and falsity) but because he did believe that some memories were unusually veridical, see Ost and Costall, "Misremembering Bartlett," 243–55.

60 David Bloor, personal communication, July 2010.

61 David Bloor, "Remember the Strong Programme?" *Science, Technology, and Human Values* 22, no. 3 (Summer 1997): 373–85.

Notes to Chapter Ten

1 Peter Freyd, personal communication, February 2011.

2 Elizabeth F. Loftus and Katherine Ketcham, *Witness for the Defense: The Accused, the Eyewitness, and the Expert Who Puts Memory on Trial* (New York: St. Martin's Press, 1991).

3 E. F. Loftus, S. Polonsky, and M. T. Fullilove, "Memories of Childhood Sexual Abuse: Remembering and Repressing," *Psychology of Women Quarterly* 18 (1994): 67–68; Elizabeth Loftus and Katherine Ketcham, *The Myth of Repressed Memory: False Memories and Allegations of Childhood Sexual Abuse* (New York: St. Martin's Press, 1994).

4 Anastasia Toufexis, "When Can Memories Be Trusted?" *Time*, October 28, 1991, 86.

5 Until the twentieth century, when its psychological meaning took shape, "confabulation" was commonly defined as "talking together," conversation, or conference (as in "confab").

6 This account comes from a personal communication with Peter Freyd, 2003. Founding members of the FMSF included Robyn Dawes (Carnegie Mellon), George Ganaway (Beth Israel Hospital, Harvard Medical School), Rochel Gelman (UCLA), Henry Gleitman (University of Pennsylvania), Lila Gleitman (Univer-

sity of Pennsylvania), Ernest Hilgard (Stanford), Philip Holzman (Harvard), John Kihlstrom (University of Arizona), Harold Lief (University of Pennsylvania), Elizabeth Loftus (University of Washington), Paul McHugh (Johns Hopkins), Ulric Neisser (Emory), Martin Orne (University of Pennsylvania), and Margaret Singer, (University of California, Berkeley).

7   "Has your grown child falsely accused you as a consequence of repressed 'memories'?" You are not alone. Please help us document the scope of this problem. Contact: 1-800-568-8882." FMSF newsletter, December 1991.

8   FMSF newsletter, December 1991.

9   This was the theme of the regional meeting spanning Virginia, West Virginia, and Washington, DC; FMSF meeting, August 21, 1993, FMSF newsletter, March 1993.

10 Peter Freyd, personal communication, August 26, 2010.

11 See Steven Epstein, *Impure Science: AIDS, Activism, and the Politics of Knowledge* (Berkeley: University of California Press, 1996).

12 "Have you been falsely accused on the basis of recovered 'memories'?" You are not alone. Help us document the extent of this problem. Contact: False Memory Syndrome Foundation through the Institute for Psychological Therapies, 1-800-568-8882." FMSF newsletter, March–April 1992.

13 FMSF newsletter, June 1992.

14 FMSF newsletter, August 1992.

15 FMSF newsletter, September 1992.

16 FMSF newsletter, August–September 1992.

17 FMSF newsletter, November 1992.

18 FMSF newsletter, July 1992: "This year there have been articles in the Los Angeles Times, Philadelphia Inquirer, Toronto Star, Miami Herald, Utah County Journal, Cleveland Plain Dealer, Kansas City Star and many other papers that discussed these issues.... False memory syndrome is now a subject that can be discussed and examined."

19 Pamela Freyd, writing in FMSF newsletter, August 1992.

20 "Seventeen percent of the people in our family counts have been threatened or actually sued. "We won but we lost," is the way some families describe their experience.... 'Meeting with your child's therapist' booklet is at the printers and will be mailed separately to FMSF members." FMSF newsletter, January 1993.

21 John E. B. Myers, "Expert Testimony Describing Psychological Syndromes," *Pacific Law Journal* 24 (1993): 1449–64. Myers noted the confusion that had arisen in court cases over such terms as "rape trauma syndrome" and "child sexual abuse accommodation syndrome" that, unlike "battered child syndrome," are commonly understood to have far less diagnostic value. Seymour Halleck et al., "The Use of Psychiatric Diagnoses in the Legal Process; Task Force Report of the American Psychiatric Association," *Bulletin of the American Academy of Psychiatry and Law* 20, no. 4 (1992): 481–99; Gayle Hanson, "Total Recall versus Tricks of the Mind," *Insight Magazine*, May 24, 1993.

22 FMSF newsletter, February 1993.

23 Ibid.

24 Ibid., January 1992.

25 "It is critical that allegations are investigated promptly, thoroughly and objectively by trained law enforcement and other professionals. . . . If these procedures had been followed we would not be getting calls from families." "Update," FMSF newsletter, July 1992.

26 Hollida Wakefield, much less active in the interview than Underwager, said she worried about the negative societal view of pedophilia, a view that might harm children.

27 "Controversial Article," FMSF newsletter, July 1993, 7. On December 19, 1993, the *London Sunday Times* reported that Underwager had appeared in England to testify for a man denied custody of his children because of child abuse allegations. In reference to his *Paidika* interview, the *Times* reported that Underwager denied ever condoning sex with children, though he said "scientific evidence" shows that "60% of women sexually abused as children reported that the experience was good for them." Underwager contended that the same could be true for boys.

28 Moira Johnston, *Spectral Evidence: The Ramona Case; Incest, Memory, and Truth on Trial in the Napa Valley* (Boston: Houghton Mifflin, 1997).

29 B. Drummond Ayers, "Father Who Fought 'Memory Therapy' Wins Damage Suit," *New York Times*, May 14, 1994.

30 Ibid.

31 Jane Gross, "'Memory Therapy' on Trial: Healing or Hokum?" *New York Times*, April 8, 1994. Gross quoted Renee Fredrickson, a psychologist and author who endorsed repressed memory treatment, as saying she hoped education for therapists would improve: "Of course, we're turning out people who don't know what they're doing," she said. "We're teaching each other or getting trained at weekend dog and pony shows."

32 E.g., *Hungerford v. Jones*, 722 A.2d 478 (N.H. 1998), and *Sawyer v. Midelfort*, 595 NW 2d 423 (Wis. 1999); "Parents Win Suit against Psychiatrist in Sex Case," *New York Times*, December 17, 1994, 9.

33 Associated Press, "Man Released after Being Held 6 Years in Repressed Memory Case," *Los Angeles Times*, July 4, 1996.

34 FMSF newsletter, August 1993.

35 "Memory and Reality: Emerging Crisis," conference program, FMSF Archives.

36 Affidavit of Richard Ofshe, December 28, 1990, First Judicial District Court, County of Santa Fe, New Mexico, *Mark Baker v. Yogi Bhajan et al.*, Case No. SF 99–2286 (C).

37 George K. Ganaway, "Alternative Hypotheses regarding Satanic Ritual Abuse Memories," paper presented at the annual convention of the American Psychological Association, San Francisco, August 1991.

38 Loftus and Ketcham, *Myth of Repressed Memory*, 89, 90, 92.

39 Ibid., 91.

40 Elizabeth F. Loftus, "The Reality of Repressed Memories," paper presented at the annual conference of the American Psychological Association, August 14–18, 1992, cassette recording no. 92-013 (Washington, DC: American Psychological Association); R. B. Klein, "The Nature of Memory: An Interview with Prof. Elizabeth F. Loftus, Ph.D.," *Verdicts, Settlements, and Tactics* 15, no. 6 (1995): 191–95.

41 Loftus and Ketcham, *Myth of Repressed Memory*, 95.

42 Associated Press, "Analyst Doubts Abuse "Memories," *Tri-City Herald*, August 14, 1992, A5; J. Morrison, "You Must Remember This," *George* 1, no. 10 (December 1996): 52; E. F. Loftus and J. E. Pickrell, "The Formation of False Memories," *Psychiatric Annals* 25 (1995): 720–25.

43 Loftus and Pickrell, "Formation of False Memories," 722.

44 Lynn S. Crook, "'Lost in a Shopping Mall': A Breach of Professional Ethics," *Ethics and Behavior* 9, no. 1 (1999): 39–50; J. A. Coan, "'Lost in a Shopping Mall': An Experience with Controversial Research," *Ethics and Behavior* 7 (1997): 271–84.

45 William Saletan, "The Memory Doctor," *Slate*, posted Friday, June 4, 2010, at 5:45 p.m. EDT; quotation is from George Orwell, *1984.*

46 Dariol M. Sacchi, Franca Agnoli, and Elizabeth Loftus, "Changing History: Doctored Photographs Affect Memory for Past Public Events," Applied Cognitive Psychology 21 (2007): 1005–22.

47 Jeffrey Masson, *The Assault on Truth: Freud's Suppression of the Seduction Theory* (New York: Farrar, Straus and Giroux, 1984); Jeffrey Masson, *Against Therapy: Emotional Tyranny and the Myth of Psychological Healing* (New York: Atheneum, 1988); Jeffrey Masson, *Final Analysis: The Making and Unmaking of a Psychoanalyst* (Reading, MA: Addison-Wesley, 1990).

48 Frederick C. Crews and Matthew H. Erdelyi, "Freud and Memory: An Exchange," *New York Review of Books*, March 23, 1995; Frederick C. Crews et al., *The Memory Wars: Freud's Legacy in Dispute* (New York: New York Review of Books, 1997); F. C. Crews, "The FMSF Scientific and Professional Advisory Board-Profiles: Frederick C. Crews, False Memory Syndrome Foundation, www.fmsfonline.org/advboard.html. For other discussions of repression in the context of the debate over false memory see J. L. Hudson and H. G. Pope, "Can Memories of Childhood Abuse Be Repressed?" *Psychological Medicine* 25 (1995): 121–26; August Piper, "What Science Says—and Doesn't Say—about Repressed Memories: A Critique of Scheflin and Brown," *Psychiatry and the Law* 25 (1997): 615–38.

49 In 1993 Ofshe predicted that recovery therapy would soon be seen as "the quackery of the twentieth century." Leon Jaroff, "Lies of the Mind," *Time Magazine*, November 29, 1993.

50 S. Duggal and L. A. Stroufe, "Recovered Memory of Childhood Sexual Trauma: A Documented Case from a Longitudinal Study," *Journal of Traumatic Stress* 11, no. 2 (1998): 301–21.

51 C. Bagley, "The Prevalence and Mental Health Sequels of Child Sexual Abuse in Community Sample of Women Aged 18 to 27," in Bagley, *Child Sexual Abuse and Mental Health in Adolescents and Adults: British and Canadian Perspectives* (Aldershot, UK: Avebury, 1995). See also J. A. Chu, L. M. Frey, B. L. Ganzel, and J. A. Matthews, "Memories of Childhood Abuse: Dissociation, Amnesia, and Corroboration," *American Journal of Psychiatry* 156, no. 5 (1999): 749–55; L. M. Williams, "Recovered Memories of Abuse in Women with Documented Child Sexual Victimization Histories," *Journal of Traumatic Stress* 8, no. 4 (1995): 649–73; C. Widom and R. Shepard, "Accuracy of Adult Recollections of Childhood Victimization: Part 1," *Psychological Assessment* 8, no. 4 (1996): 412–21; C. Widom and R. Shepard, "Accuracy of Adult Recollections of Childhood Victimization: Part 2, Childhood

Sexual Abuse," *Psychological Assessment* 9 (1997): 34–46. For a more extensive discussion of the relevant literature see Daniel P. Brown, D. Corydon Hammond, and Alan W. Scheflin, *Memory, Trauma Treatment, and the Law* (New York: W. W. Norton, 1998).

52 D. Corwin and E. Olafson, "Videotaped Discovery of a Reportedly Unrecallable Memory of Child Sexual Abuse: Comparison with a Childhood Interview Taped 11 Years Before," *Child Maltreatment* 2, no. 2(1997): 91–112.

53 E. F. Loftus and M. J. Guyer, "Who Abused Jane Doe? The Hazards of the Single Case History: Part 2," *Skeptical Inquirer* 26, no. 4 (2002): 37–40.

54 Carol Tavris, "Whatever Happened to 'Jane Doe'?" *Skeptical Inquirer* 32, no. 1 (2008): 28–30.

55 Richard Ofshe and Ethan Watters, *Making Monsters: False Memories, Psychotherapy, and Sexual Hysteria* (New York: Scribner's, 1995); Richard Ofshe and Ethan Watters, *Therapy's Delusions: The Myth of the Unconscious and the Exploitation of Today's Walking Worried* (New York: Scribner's 1995); Margaret Thaler Singer and Richard Ofshe, "Thought Reform Programs and the Production of Psychiatric Casualties," *Psychiatric Annals* 20, no. 4 (April 1990): 188–93; Richard Ofshe and Margaret T. Singer, "Attacks on Peripheral versus Central Elements of Self and the Impact of Thought Reforming Techniques," *Cultic Studies Journal* 3, no. 1 (1986): 3–24; Charlotte Allen, "Brainwashed! Scholars of Cults Accuse Each Other of Bad Faith," *Lingua Franca*, December 1998, 26–37; Lawrence Wright, *Remembering Satan: A Case of Recovered Memory and the Shattering of an American Family* (New York: Knopf, 1994), 134–46; K. Olio and W. Cornell, "The Facade of Scientific Documentation: A Case Study of Richard Ofshe's Analysis of the Paul Ingram Case," *Psychology, Public Policy, and Law* 4, no. 4 (1998): 1182–97; Ethan Watters, "The Devil in Mr. Ingram," *Mother Jones* 16, no. 4 (1991): 30–68.

56 American Psychiatric Association, *Statement on Memories of Sexual Abuse* (Washington, DC: American Psychiatric Association, 1993); American Psychological Association, *Questions and Answers about Memories of Childhood Abuse* (Washington, DC: American Psychological Association, 1995); American Medical Association, Council on Scientific Affairs, *Memories of Childhood Sexual Abuse* (Washington DC: American Medical Association, 1994).

57 "Final Report of the APA Working Group on Investigation of Memories of Childhood Abuse," *Psychology, Public Policy, and Law* 4, no. 4 (December 1998): 933–40; Conclusions of the APA Working Group on Investigations of Memories of Childhood Abuse, www.apa.org/pi/memories_report/section7.pdf; C. Brook Brenneis, "Final Report of the APA Working Group on Investigation of Memories of Childhood Abuse: A Critical Commentary," *Psychoanalytic Psychology* 14 (1997): 531–47; J. L. Alpert, L. S. Brown, and C. A. Courtois, "Nonissues, Real Issues, and Bias: Comments on C. Brooks Brenneis's Commentary," *Psychoanalytic Psychology* 16 (1999): 94–102.

58 See D. Brown, A. Scheflin, and C. L. Whitfield, "Recovered Memories: The Current Weight of the Evidence in Science and in the Courts," *Journal of Psychiatry and Law* 27 (Spring 1999): 5–156.

59 The records now have a permanent archive at the Center for Inquiry in Hamburg, New York, funded by a financial gift from one of Peter Freyd's students.

60 For a late and arguably minor example, see Frederick Crews, *Postmodern Pooh* (New York: North Point Press, 2002), 126.

Notes to Chapter Eleven

1   Hagar Gelbard-Sagiv, Roy Mukamel, Michal Harel, Rafael Malach, and Itzhak Fried, "Internally Generated Reactivation of Single Neurons in Human Hippocampus during Free Recall," *Science* 322, no. 5898 (October 3, 2008): 96–101. See also Greg Miller, "Neuroscience: Hippocampal Firing Patterns Linked to Memory Recall," *Science* 321, no. 5894 (September 5, 2008): 1280b; A. J. Silva, Y. Zhou, T. Rogerson, J. Shobe, and J. Balaji, "Molecular and Cellular Approaches to Memory Allocation in Neural Circuits," *Science* 326, no. 5951 (October 16, 2009): 391–95; L. R. Squire, "Memory and Brain Systems: 1969–2009," *Journal of Neuroscience* 29 (2009): 12711–16; P. R. Shirvalkar, "Hippocampal Neural Assemblies and Conscious Remembering, *Journal of Neurophysiology* 101 (2009): 2197–200; "The Neuroscientist Comments," *Neuroscientist* 15 (April 2009): 122–23; "Perspectives on Neuroscience and Behavior," *Neuroscientist* 14, no. 5 (2009): 124–25; H. Lehn, H.-A. Steffenbach, N. M. Van Strien, D. J. Veltman, M. P. Witter, and A. K. Haberg, "A Specific Role of the Human Hippocampus in Recall of Temporal Sequences," *Journal of Neuroscience* 29 (2009): 3475–84; M. P. Karlsson and L. M. Frank, "Network Dynamics Underlying the Formation of Sparse, Informative Representations in the Hippocampus," *Journal of Neuroscience* 28 (2008): 14271–81.

2   Gelbard-Sagiv et al., "Internally Generated Reactivation."

3   James McGaugh, "Memory—a Century of Consolidation," *Science* 287, no. 5451 (2000): 249.

4   On theories of psychological trauma as "sciences of the literal," see Ruth Leys, *Trauma: A Genealogy* (Chicago: University of Chicago Press, 2000), 229–65. And on theories of the neurophysiology of traumatic memory and the role of stress hormones in such theories, see Alan Young, *Harmony of Illusions: Inventing Posttraumatic Stress Disorder* (Princeton, NJ: Princeton University Press, 1995), 274–77.

5   Larry Cahill and James L. McGaugh, "Mechanisms of Emotional Arousal and Lasting Declarative Memory," *Trends in Neurosciences* 21, no. 7 (1998): 294–99.

6   Lennart Levi, ed., *Emotions, Their Parameters and Measurement* (New York: Raven Press, 1975), 209–34.

7   Sven-Åke Christianson, ed., *The Handbook of Emotion and Memory: Research and Theory* (Hillsdale, NJ: Erlbaum, 1992), 245–68; B. Bohus, "Humoral Modulation of Learning and Memory Processes," in *The Memory System of the Brain*, ed. J. Delacour (Singapore: World Scientific, 1994), 338–64; B. R. Kaada, "Stimulation and Regional Ablation of the Amygdaloid Complex with reference to Functional Representation," in *The Neurobiology of the Amygdala*, ed. B. Eleftheriou (New York: Plenum Press, 1972), 205–82; G. V. Goddard, "Amygdaloid Stimulation and Learning in the Rat," *Journal of Comparative Physiology and Psychology* 58 (1964): 23–30, 40; Larry Cahill and James L. McGaugh, "Amygdaloid Complex Lesions Differentially Affect Retention of Tasks Using Appetitive and Aversive Reinforcement," *Behavioral Neuroscience* 104 (1990): 532–43; Larry Cahill and James L. McGaugh,

"Mechanisms of Emotional Arousal and Lasting Declarative Memory," *Trends in Neurosciences* 21 (1998): 294–99.

8 Benno Roozendaal and James L. McGaugh, "Amygdaloid Nuclei Lesions Differentially Affect Glucocorticoid Induced Memory Enhancement in an Inhibitory Avoidance Task," *Neurobiology of Learning and Memory* 65 (1996): 1.

9 P. C. Holland and M. Gallagher, "Amygdala Circuitry in Attentional and Representational Processes," *Trends in Cognitive Science* 3 (1999): 65–67; N. M. White and R. J. McDonald, "Multiple Parallel Memory Systems in the Brain of the Rat," *Neurobiology of Learning and Memory* 77 (2002): 125–84; A. Easton and D. Gaffan, "The Amygdala and the Memory of Reward: The Importance of Fibres of Passage from the Basal Forebrain," in *The Amygdala: A Functional Analysis*, ed. J. P. Aggleton, 2nd ed. (New York: Oxford University Press, 2000): 569–86; Mark G. Baxter and Elisabeth A. Murray, "Reinterpreting the Behavioral Effects of Amygdala Lesions in Non-human Primates," in *The Amygdala: A Functional Analysis*, ed. J. P. Aggleton, 2nd ed. (New York: Oxford University Press, 2000), 569–86.

10 J. R. Misanin, R. R. Miller, and D. J. Lewis, "Retrograde Amnesia Produced by Electroconvulsive Shock after Reactivation of a Consolidated Memory Trace," *Science* 160, no. 3827 (May 3, 1968): 554–55; R. G. Dawson and J. L. McGaugh, "Electroconvulsive Shock Effects on a Reactivated Memory Trace: Further Examination," *Science* 166 (1969): 525–27; Natalie C. Tronson and Jane R. Taylor, "Molecular Mechanisms of Memory Reconsolidation," *Nature Reviews Neuroscience* 8 (April 2007): 262–75.

11 Cahill and McGaugh, "Mechanisms of Emotional Arousal and Lasting Declarative Memory."

12 Glenn E. Schafe and Joseph E. Ledoux, "Memory Consolidation of Auditory Pavlovian Fear Conditioning Requires Protein Kinase A in the Amygdala," *Neuroscience* 20 (2000): 1–5.

13 Other experiments found no such relationship. A. H. van Stegeren, W. Everaerd, and L. J. Gooren, "The Effect of Beta-Adrenergic Blockade after Encoding on Memory of an Emotional Event," *Psychopharmacology* (Berlin) 163, no. 2 (September 2002): 202–12.

14 Michael C. Anderson, Kevin N. Ochsner, Brice Kuhl, Jeffrey Cooper, Elaine Robertson, Susan W. Gabrieli, Gary H. Glover, and John D. E. Gabrieli, "Neural Systems Underlying the Suppression of Unwanted Memories," *Science* 303, no. 5655 (2004): 232–35.

15 Quoted in Scott LaFee, "Blanks for the Memories: Someday You May Be Able to Take a Pill to Forget Painful Recollections," *San Diego Union Tribune*, February 11, 2004.

16 Peter D. Kramer, *Listening to Prozac: A Psychiatrist Explores Antidepressant Drugs and the Remaking of the Self* (New York: Viking, 1993).

17 For a more sustained discussion of this film, in the context of the representation of memory in popular film, see Fernando Vidal, "Memory Movies and the Brain," in *The Memory Process: Neuroscientific and Humanistic Perspectives*, ed. Suzanne Nalbantian, Paul Matthews, and James L. McClelland (Cambridge, MA: MIT Press, 2010).

18 Steven Johnson, "The Science of Eternal Sunshine: You Can't Erase Your Boy-

friend from Your Brain, but the Movie Gets the Rest of It Right," *Slate*, posted Monday, March 22, 2004, at 4:33 p.m. EST.

19 "Beyond Therapy: Biotechnology and the Pursuit of Happiness," report of the President's Council on Bioethics, Washington, DC, October 2003.

20 Tara Mckelvey, "Combat Fatigue," *American Prospect* 19, no. 7 (July–August 2008): 5.

21 See also Gilbert Meilaender, "Why Remember?" *First Things* 135 (2003): 20; Greg Miller, "Learning to Forget," *Science* 304, no. 5667 (2004): 34; Eric Board, "The Guilt Free Soldier," *Village Voice*, January 22, 2003; Sharon Beglery, "A Spotless Mind May Ease Suffering, but Erase Identity," *Wall Street Journal*, August 19, 2005; Rob Stein, "Is Every Memory Worth Keeping?" *Washington Post*, October 19, 2004, A1.

22 For a discussion of the (optimistic and alarmist) claims currently being made about the significance of the neurosciences for our understanding of ourselves and how we manage our own behavior, see Fernando Vidal, "Brainhood, Anthropological Figure of Modernity," *History of the Human Sciences* 22, no. 1 (2009): 5–36.

23 For an intellectual history of the "consubstantiality" of memory and identity since the eighteenth century see Vidal, "Memory, Movies and the Brain."

INDEX

Page numbers in italics refer to illustrations.

abreaction, 54–55, 56, 57, 88, 194
Actors Studio, 99
age regression, 104, 123, 149
Ahmadinejad, Mahmoud, 247
Alcoholics Anonymous, 245
alien abduction, 123
altered memory, 16–17. *See also* false memory
Alzheimer's disease, 270
American Civil Liberties Union, 183
American Heart Association, 229
American Institute of Hypnosis, 129
American Medical Association, 52, 126, 251
American Office of Strategic Services (OSS), 71
American Photographic Annual, 94
American Psychiatric Association, 248, 251
American Psychological Association (APA), 245, 251–52
amnesia, 53–54, 56–57, 67, 175. *See also* recovered memory; repressed memory
amygdala, 85, 86, 263–64, 265
Amytal. *See* sodium amytal
Ancient Wisdom, 112
Anderson, Michael, 266
anesthesia, 271–72
anthropology, 200–201, 209–10
APA. *See* American Psychological Association (APA)
Arons, Harry, 139–41, 143, 145

Assassination Science, 170
Atlantic, 31
Atwell, Henry, 109–10
authenticity, 3, 257–61, 270–71
automatism, 86–87

Bailey, F. Lee, 128–29, 135
barbiturates. *See* drugs
Barker, William J., 106, 114, 115
Barr, Roseanne, 193
Barthes, Roland, 166, 295n16
Bartlett, Frederic, 198–223; and clinical psychology, 212–13; and collective memory, 215–16; and constructive memory, 45, 82–83, 175, 198, 221, 258; and conventionalization, 201–2, 203, 212, 214, 215, 216, 219, 222; early life and career, 200–202; influence of, 220–23; and meaning, 199, 203–12; and perception, 199, 202–13; and psychoanalysis, 212–13, 214–15; *Psychology and Primitive Culture,* 215; *Remembering,* 199, 214, 215, 218, 222; reproduction experiments, 6, 202–12, *203, 204, 207;* and schema, 216–20; and thinking, 217, 222; "The War of the Ghosts," 207–12, 214, 219
Barton, Patti, 187
Bass, Ellen, 181
Battle Exhaustion, 62
battle trauma, 53, 55–57, 59–69, 269. *See also* post-traumatic stress disorder (PTSD)
Beecher, Henry, 65–66

behaviorism, 45

Belli, Melvin, 128, 129, 136, 139

Bernstein, Morey: family background, 104; hypnotism of Virginia Tighe, 103, 105–6, 119–20; interest in age regression, 104; LP of hypnotism session, 107; readers' letters to, 111–14; *The Search for Bridey Murphy,* 103, 106–7, 108, 115, 121. *See also* Bridey Murphy affair

Bianchi, Kevin, 151–52

Biden, Joseph, 191–92

Binet, Alfred, 30, 206

biofeedback theories, 167

Birren, James, 140

Blade Runner, 98

Blake, George, 133

blocked memory. *See* repressed memory

Bloor, David, 221–22

Boas, Franz, 208–9

body memory, 189

Boston Strangler, 58, 135–36

Bowerman, R., 113, 287n37

Bradshaw, John, 190

Braid, James, 130

brain, 75–93, 97, 99–102, 139, 258–61

brain maps, *76,* 77, 82

brainwashing, 58, 127–28, 131–34, 281n16

Bravos, Zachary, 240, 241

Bridey Murphy affair, 4, 7, 103–23, 258, 271; details given by Tighe under hypnotism, 105–6, 110–11; film about, 107, 121–22; and history, 104; investigations of, 114–20; Irish American perspective on, 110; LP of hypnotism session, 107; press interest in, 106–10, 114–21; and psychiatry, 120; and psychology, 116–17; public response to, 107–8, 111–14; and religion, 109–10; salon culture, 107–8, *108, 109; The Search for Bridey Murphy,* 103, 106–7, 108, 115, 121

British Journal of Psychology, 201

Brown, Roger, 157–60, 170, 171–73, 174, 176, 178, 263, 266

Brunner, Ray, 117

Bryan, William Jennings, 127

Bryan, William Joseph, 127–31, 133–38, 141, 144, 145

Bubbles, Mr., 239

Buckhout, Robert, 197, 198

Buckley, William, 108–9

Bureau of American Ethnology Bulletin, 208

Burt, Cyril, 200

Busch, Henry Adolphe, 134–35, 290n24

Bush, George W., 248, 268

Bush, Vannevar, 54

California Association of Marriage and Family Therapists, 240–41

California Defense Council, 188

cameras. *See* photography

Cameron, D. Ewan, 131–32

Carmichael, Leonard, 134

Cartier-Bresson, Henri, 162, 294n10; *The Decisive Moment,* 162

Caruth, Cathy, 178

Case Western Reserve University, 145, *146*

Cathlamet. *See* Kathlamet

Catholic Church child abuse, 253

Cattell, J. M., 45

Cayce, Edgar, 105, 112, 123

cell assembly, 83

Century of Progress Exposition, 50

Cetus, 244

Challenge of Reincarnation, The (Luntz), 112

Challenger space shuttle, 173–74

Chandler, Sharon S., 241

character, 266–67. *See also* self

Chicago American, 117, 118, 119–20

Chicago Defender, 108

Chicago Tribune, 48

Child, Julia, 158

child abuse. *See* sexual abuse

China, 105

Chowchilla kidnapping, 147–48, 177

Christison, J. Sanderson, 11–15, 19

CIA, 127

Cinemascope, 94

civil liberties, 48

Claparède, Édouard, 30

Clegg, J. L., 57

Clemens, Roger, 248

Clinton, Bill, 247

cognition, 209

cognitive psychology, 101, 220–21, 258

Cognitive Psychology (Neisser), 220–21

cold war, 127, 131, 133–34

collective memory, 158–60, 215–16, 241–42, 254

Combat Exhaustion, 63–64, 66, 67, 69

Committee for Skeptical Inquiry (CSI), 254

Committee for the Scientific Investigation of Claims of the Paranormal (CSICOP), 254

communism, 105, 127, 131, 133–34

complexity, 222

Conditional Reflex Therapy (Salter), 132

Condon, Richard, 132

confabulation, 8, 153, 228–29, 234, 303n5

confessions: false, 9–20; and hypnosis, 135–37, 143; and truth serum, 40, 47–51

consciousness, 90–91, 92–93, 94; stream of, 78–79, 92, 97–98. *See also* unconscious memory

consolidation, 175–76, 262–66. *See also* reconsolidation

constructive imagination, 26–27

constructive memory, 45, 82–83, 175, 198, 221, 257–58

context, 200, 215–16, 219, 222

contingency, 215

conventionalization, 201–2, 203, 212, 214, 215, 216, 219, 222

Corcoran, Bill, 184

Cornell University, 174

Cornell v. Superior Court of San Diego, 289n2

cortisol, 262, 265

Corwin, David, 250

Courage to Heal, The, 191, 235, 236

Cox, Anne, 191, 299n36

Crews, Frederick, 248

crime, increase in, 37

Cultee, Charles, 208–9

culture and memory differences, 301n32

Daily News, 117

Dalai Lama, 105

Davis, Laura, 181

Decisive Moment, The (Cartier-Bresson), 162

demand characteristics, 149–51, 196, 228, 248

Dementia 13, 141, *142*

Denver Post, 106, 117, 118

Department of Child and Family Services, 236

DeSalvo, Albert, 134–36

De Vilbias, John A., 41, 42

Diagnostic and Statistical Manual, 248

Diamond, Bernard, 154–55

Dick, Philip K.: *Do Androids Dream of Electric Sheep?,* 98

Dick Tracy comic strip, *48*

Dilbert comic strip, *242*

Do Androids Dream of Electric Sheep? (Dick), 98

Dollard, John, 116

Donovan's Brain, 139, *140*

Doonesbury comic strip, *241*

drugs: ether, 57–58, 112; and malingerers, 69–70; and memory editing, 261–72; Nembutal, 63; propranolol, 265–66, 269, 270; Prozac, 267;

scopolamine, 35–52, 276n6; sodium amytal, 35, 53, 54–55, 55–58, 63, 69–70, 239, 280n7; Sodium Pentothal, 35, 53, 59–73, 89–90, 126, 137; tetrahydrocannabinol acetate, 71; and witness testimony, 270. *See also* truth serum

DSM. *See Diagnostic and Statistical Manual*

Dudum, Tom, 240

Duggan, Pat, 122

Eastman Kodak. *See* Kodak

Ebbinghaus, Hermann, 200

electroconvulsive shock therapy (ECT), 264

Elliotson, John, 130

emotion: and flashbulb memories, 157–60, 171, 173, 174, 175–76, 263; and memory, 61, 261–70

engram, 83

Epilepsy and the Functional Anatomy of the Human Brain (Penfield and Jasper), 97

epilepsy research, 75–77, 84–85, 86, 258

epinephrine, 262

Erickson, Milton, 120

Eternal Sunshine of the Spotless Mind, 267–68

ether, 57–58, 112

eugenics, 28

evolutionary theory, 35, 170–71

experimenter effect, 148–49. *See also* demand characteristics

experts and non-experts, 6–7, 80–82, 104, 116, 120, 190–91; experts and counter-experts, 193

Fabing, Howard, 65, 282n37

false confessions, 9–20

false memory, 225–55; and drugs, 44; false memory syndrome, 225–26, 230, 244–45; and the law, 9–18, 24, 235–36, 239–43, 250–51; research on, 243–50; and sexual abuse, 225–45, 249–55

False Memory Syndrome Foundation (FMSF), 227, 229–39, 243–45, 253–55

Far Side, The, 172

FBI, 153, 168

Fechner, Gustav, 200

Feindel, William, 86–87

feminist movement, 181, 183, 190–91

film: and brainwashing, 131–34; Cinemascope, 94; flashbacks, 64–65, 94; home movies, 93–94, 160–66; and hypnosis, 129–30; as an impulse for murder, 134–35; and memory, 25–26, 28–29, 54, 63–69, 82, 91–98, 132–33, 141, 160–66, 259–60; and narrative, 211; psychology of, 25–29; and scientific and medical research, 95–96; 3D, 94

flashbacks, 64–65, 93, 94, 189, 190
flashbulb memory, 157–78; *Challenger* space
shuttle, 173–74; and emotion, 157–60, 171, 173,
174, 175–76, 263; Kennedy assassination, 157,
158–61, 165, 167, 168–70; Now Print!, 166–68,
169–71, 263; and photography, 160–66, 167,
173, 174; and posttraumatic stress disorder,
176–78; and race, 159–60; reliability of, 171–74
flash of light, 16–17, 19, 28
FMSF. *See* False Memory Syndrome Foundation (FMSF)
forensic hypnosis, 125–55; and brainwashing,
127–28; confabulation, 153, 228–29; and
confession, 135–37, 143; demand characteristics, 149–51, 228, 248; devices for, 138–39;
and the FBI, 153; influence of film on
defendants, 134–35; and intent, 134–35, 144;
lay hypnosis, 139–41, 143; legitimacy of, 145,
146–47; and the military, 145, 292n46; and the
police, 126–27, 137, 139–40, 145–47, 148, 153, 155,
292n46; and recovered memory, 194; and social psychology, 148; and suggestion, 148–55;
training in, 126–27, 129, 137–38, 145–46, *146*,
148, 292n53; and trauma, 176–77; unreliability
of, 148–55; and witness testimony, 129, 143–55
forensic psychology, 6, 221, 257–58
forensic science, 48–49
Forum, 51
Frankenheimer, John, 131, 132
Franklin, George, 179–80, 192–93, 227, 242–43,
250
Franklin, Robert Walker, 42
Fredrickson, Renee, 305n31
Freud, Sigmund, 29, 35, 59, 212–13, 248; *Psychopathology of Everyday Life,* 212
Freyd, Pamela, 226, 227–43, 236–37, 238, 243–54
Freyd, Peter, 226, 227–32, 238
Frye ruling, 40, 42, 49
functional localization, 77, 81–82

Gabrieli, John, 266
Genentech, 244
Genetic Alliance, 245
Gilbert, Melissa, 184
Goddard, Calvin, 49
Graham, Billy, 158
Grinker, Roy, 59–63, 65, 71–73, 126; *Men Under
Stress,* 63, 68–69
Gross, Hans, 30
Gross, Jane, 305n31

Haddon, Alfred Court, 201, 216
Haeckel, Ernst, 35
Haines, William H., 71–72
hallucination *vs.* memory, 77–78, 82, 101
Harris, Elihu, 185, 298n14
Harvard Women's Law Journal, 183
Haywood, William, 21–24
Head, Henry, 45, 201, 212, 216–18, 300n6
Hearst, Patty, 152
Heath, Simon, 166
Heaven and Earth Magic, 97, 98
Hebb, Donald, 83–84; *The Organization of
Behavior,* 83
Heirens, William, 71–72
Herman, Judith, 181
Herman, Morris, 54
high fidelity, 94, 96
Hill, Ernie, 115
Hillside Strangler, 151–52
hippocampus, 85, 86, 89, 101, 258, 263
Hitchcock, Alfred, 135
Hollister, Elizabeth Martha, 9–11
home movies, 93–94, 160–66. *See also* film
Horizon, 151–52
Horsley, John Stephen, 54–55, 56, 57, 59, 280n4
House, Robert Ernest, 33, 36–44, *38,* 45–47, 48,
51, 52, 54, 199
Hubbard, L. Ron, 290n21
Hudson, George, 42, 43
Hulbert, Martin, 41
Humes, James, 168–69
Hunter, Edward, 132
Hurd, State v., 154
Hurlbut, William B., 267
Huston, John, 66–69
hypnosis: acceptance of by AMA, 126; age
regression, 104, 149; and battle trauma,
53, 66–67; and false confessions, 9, 11;
hypnoscope, 15, *16*; lay hypnosis, 139–41,
143; mechanisms of, 139; and memory, 113,
116; and past lives, 4, 7, 103–21; and religion,
111; salon culture, 107–8, *108, 109. See also*
forensic hypnosis

Illinois Law Review, 29
Inbau, Frederick, 44, 49, 50
incest. *See* sexual abuse
Ingram, Paul, 251
inkblots, 206
inner child, 190

Instinct and the Unconscious (Rivers), 212
Institute for Psychological Therapies, 228, 237
intent, 134–35, 144
Isabella, Marche, 239–41
Ivens, Richard, 9–20, 28

James, William, 13, 15; *Principles of Psychology,* 92
Jasper, Herbert, 80; *Epilepsy and the Functional Anatomy of the Human Brain,* 97
Jester, Perry, 111
Johns Hopkins University, 173–74
Johnson, Irene Smith, 113–14, 288n38
Journal of the American Institute of Hypnosis, 141
Jung, Carl, 215

Kaden, Bernard, 120
Karloff, Boris, 141
Karney, Shari, 183–84, 187
Kathlamet, 209
Kaufman, M. Ralph, 65
Keeler, Leonard, 49, 50
Kennedy, John F., assassination of, 133, 157, 158–61, 165, 167, 168–70
King, Martin Luther, Jr., 159
Kinkead, Eugene, 132
Kirchway, George W., 46
Klehs, Johan, 184, 185
Klein, Charles, 31; *The Lion and the Mouse,* 18; *The Music Master,* 18; *The Third Degree,* 18–20, *19*
Kline, Milton, 120
Kodak, *94, 95, 95,* 161–65, *163, 164,* 174; Kodak moment, 162–64
Korean War, 131
Kris, Ernest, 91
Kroger, William S., 147–48
Krystal, Phyllis, 112
Kubie, Lawrence, 88–91
Kulik, James, 157–60, 170, 171–73, 174, 176, 178, 263, 266

Lambert, Carl, 69
Land, Edwin, 166
Langley, Noel, 121–22
Larson, Gary, 171, *172*
Lashley, Karl, 83
law, 5–6; and false memory, 235–36, 239–43, 250–51; and psychology, 9–10, 11–20, 21–25, 29–32, 147; and psychotherapy, 239–43; and recovered memory, 5, 179–96; and

sexual abuse, 182–88, 235–36, 239–43, 250–51; statutes of limitations, 5, 182–88; and truth serum, 36–43, 45–51, 71–73. *See also* forensic hypnosis; witness testimony
lay hypnosis, 139–41, 143
learning, synaptic account of, 83
Ledoux, Joseph, 264–65
Lenin, Vladimir, 134
Leopold, Nathan, 46
Lerner, Alan J., 123
Let There Be Light, 66–69, *68*
Levendula, Dezso, 145
Lévy-Bruhl, Lucien, 209
Leys, Ruth, 178
Liberace, 108
Lieberman, Joe, 247–48
lie-detection technologies, 37, 46. *See also* hypnosis; truth serum
Lief, Harold, 228
Life, 111, 113, 114–16, 120, 122, 160–61
light, flash of. *See* flash of light
Lilly, John, 90–91
Lion and the Mouse, The (Klein), 18
Lipsker, Eileen, 179–80, 192–93, 242
literal memories, 178
Livingston, Robert B., 166–70, 263. *See also* Now Print!
Loeb, Richard, 46
Loftus, Elizabeth, 101, 175, 193, 196, 198, 221, 227, 245–47, 248, 250
Los Angeles County Criminal Bar Association, 129
Los Angeles Police Department, 148
Ludwig, Alfred O., 70, 72
Luntz, Charles: *The Challenge of Reincarnation,* 112
Lyman, Ruth, 115

MacDonald, John D., 126
Malcolm X, 159
malingerers, 69–70, 71–72
Manchurian Candidate, The, 130–34, 135
Manhattan Project, 125, 134
Margolin, Sydney, 90
Marks, John, 289n14
Massachusetts General Hospital, 269
Matthews, Al, 129, 134
McCloskey, Michael, 174
McClure's Magazine, 22, 31, 37
McDougall, William, 212, 300n6

McGaugh, James, 176, 263
McGrievy, Susan, 183
McHugh, Paul, 229
McKelvey, Tara, 269
McMartin day care scandal, 184–85
McNally, Richard, 249
McVeigh, Timothy, 263
meaning, 199, 203–12
media, 3–5. *See also* film; photography; tape recording
Meeker, George, 27
Memex, 54
memory: and authenticity, 3, 257–61, 270–71; body, 189; constructive, 45, 82–83, 198, 221, 257–58, 257–78; editing of, 261–72; *vs.* hallucination, 77–78, 82, 101; literal, 178; permanence of, 5, 82–83, 93, 96, 101, 128, 220, 262, 264; reliability of, 171–74, 197–98, 271; unconscious, 34–35, 36, 43–45, 56, 91, 141–43, 271. *See also* false memory; flashbulb memory; recovered memory; repressed memory
memory cortex, 87
Memory Palace, 268
Men Under Stress, 63, 68–69
mesmerism, 34
Meyer, Max, 13
Mezer, Robert Ross, 136
Miller, Neal E., 167
Milner, Brenda, 84–87, 89–90, 100–101
MKULTRA, 127, 141
Mohr, Jane, 187
moment: Kodak, 162–64; photographic, 161–62, 166
Mondale, Walter, 68
Montreal Neurological Institute (MNI), 75, 80, *81*, 83–85, *89*
motion pictures. *See* film
Muehlberger, Clarence, 49, 50
Munk, Hermann, 216
Münsterberg, Hugo, 44, 147, 197, 199, 211, 214, 271, 275n47; experiments on color perception, *22*; and film, 25–29, 94; and the Haywood case, 21–24; home burglary of, 20, 31; influence of, 198, 258; and the Ivens case, 9–10, 13–14, 15–20, 25; *On the Witness Stand*, 24–25; *The Photoplay*, 25; public opinion of, 30–32; and Wigmore, 29–30; and word association tests, 46
Murphy, Bridey. *See* Bridey Murphy affair
Music Master, The (Klein), 18

Myers, Charles, 200, 201, 202, 212, 214
Myers, John E. B., 304n21

narcoanalysis, 53–73
narrative, 210–12; *vs.* historical truth, 186
Nason, Susan, 179–80
National Institute of Mental Health, 168
National Institute of Neurological Diseases and Blindness, 168
National Review, 108–9
Nebb, Arthur, 136–37, 291n31
Neisser, Ulric, 172–73, 174, 222; *Cognitive Psychology*, 220–21
Nembutal, 63
neural-net theory, 83, 167
neuropsychology, 83
neurosurgery, 75–101, 258–59
Newlander, Ben Ami, 183
Newlander, Lorey, 182–83, 185
Newsweek, 226
New Yorker, 243
New York Times, 18, 46, 103, 107, 135, 260
Now Print!, 166–68, 169–71, 263

Obama, Barack, 247
Ofshe, Richard, 244, 248–49, 251
Oklahoma City bombing, 263
On a Clear Day You Can See Forever, 123
On the Witness Stand (Münsterberg), 24–25
Oprah, 226, 259, 260
Orchard, Harry, 21–24
Organization of Behavior, The (Hebb), 83
Orne, Martin, 132, 148–55, 187, 196, 198, 228–29, 244, 248
Orwell, George, 247, 248
OSS, 127
Ostow, Mortimer, 91
Oswald, Lee Harvey, 133

Packard, Vance, 132
Paidika, the Journal of Paedophilia, 238
Papp, State of Ohio v., 152
Paramount-Bray Pictograph, 26–27
parapsychology, 105, 111
Parkyn, Herbert, 12, 19
past lives, 103–21; and religion, 111; salon culture, 7, 107–8, *108, 109*. *See also* Bridey Murphy affair
Payne, Shirley, 134
Pearl Harbor, 173

pedophilia, 237–39. *See also* sexual abuse

Penfield, Wilder, 75–102, *76*; brain maps, *76*, 77, 82; and cognitive psychology, 101; and collaboration, 79–82; *Epilepsy and the Functional Anatomy of the Human Brain,* 97; and epilepsy research, 75–77, 84–85; and film, 91–98, 140–41, 160, 258–60; influence on the creative arts, 96–99; and memory, 77–102; and psychoanalysis, 87–91; and psychology, 82–87, 92; relationship with his patients, 80–82; and self, 271; sketch of memory mechanism, *100*; and the theater, 98–99; and wire recording, 91

Penfield Mood Organ, 98

Pentothal. *See* Sodium Pentothal

People v. Quaglino, 152

perception, 199, 202–13

Pheaster, Hugh Macleod, 292n42

Philippe, Jean, 202

phonograph, 3

photography: digital, 174–75; Kodak, *94, 95, 95,* 161–65, *163, 164,* 174; and memory, 160–66, 167, 173, 174, 213–14; Polaroid, *165,* 166, 167, 170

Photoplay, The (Münsterberg), 25

physiology of time, 78–79

Pillemer, David, 171

Pitman, Roger, 269–70

Pliny the Elder, 34, 43

Polaroid, *165,* 166, 167, 170

police: and hypnosis, 9, 11, 126–27, 139–40, 145–47, 148, 153, 155, 292n46; and torture, 46

posttraumatic stress disorder (PTSD): definition of, 181; drug therapies for, 266, 269; and flashbulb memories, 176–78; and sexual abuse, 181; and stress hormones, 262, 263, 266. *See also* battle trauma

powerization, 133–34

Powers, Melvin, 111

practice, 219

President's Council on Bioethics, 268–69, 270, 271

Price, Vincent, 130

primitive *vs.* modern, 209–10

Prince Arthur and Hubert, 204, 205

Principles of Judicial Proof (Wigmore), 32

Principles of Psychology (James), 92

propranolol, 265–66, 269, 270

Prozac, 267

psychiatry: and the Bridey Murphy affair, 120; and hypnosis, 125–26; during World War I,

54, 212–13; during World War II, 53–54, 55–70, 73, 125, 126, 281n18, 281n21

Psycho, 134–35

psychoanalysis, 35, 55, 59–60, 87–91, 186–87, 212–13, 214–15, 248

psychology: applied, 14, 15; and the Bridey Murphy affair, 116–17; clinical, 212–13; cognitive, 101, 220–21, 258; of film, 25–29; and law, 9–10, 11–20, 21–25, 29–32, 147; and neurosurgery, 82–87; and the press, 14; social, 132, 148

Psychology and Primitive Culture (Bartlett), 215

Psychology of the Emotions (Ribot), 99

Psychopathology of Everyday Life (Freud), 212

psychotherapy, 139, 239–43

Quaglino, People v., 152

race and flashbulb memories, 159–60

Ramona, Gary, 239–43, 250

Ramona, Holly, 239–43, 250

Randall, Tony, 158

Rank, Otto, 215

Ray, Ed, 147–48

Ready, Walter, 110

Reagan, Ronald, 171

recollection, 199

reconsolidation, 264, 266, 268

reconstructive memory. *See* constructive memory

recovered memory: and the law, 179–96; reliability of, 177, 195; and sexual abuse, 180–96, 225–27, 235, 239–43, 249–54. *See also* false memory; repressed memory

Rees, W. Linford, 69

reincarnation, 105. *See also* past lives

reinforcement, 167–68

Reiser, Martin, 292n53

religion and past lives, 109–10, 111

Remembering (Bartlett), 199, 214, 215, 218, 222

repressed memory: and battle trauma, 4, 53–54, 56–57, 61, 64, 72–73; and neurosurgery, 90–91; and sexual abuse, 181, 194, 225–27, 235, 239–43, 249–54. *See also* amnesia; false memory; recovered memory

Ribot, Theodule: *Psychology of the Emotions,* 99

Richards, I. A., 210

Riemersma, Mary, 240–41

Ritchie, Jeffrey Alan, 152

Rivers, William H. R., 200, 201–2, 209–10, 212, 214, 216; *Instinct and the Unconscious,* 212

RMS *Lancastria*, 56–57
Robinson, Henry Morton, 51
Roosevelt, Edith Kermit, 134
Rorschach, Hermann, 206
Rosenthal, Robert, 132, 148
Russian Scandal, 202–3, 208

Salten, Melissa, 183
Salter, Andrew: *Conditional Reflex Therapy*, 132
Sandstrom, Richard, 148
Sargant, William, 55–58, 59, 61, 67, 126; *The Treatment of War Neuroses*, 56–57
SCDL. *See* Scientific Crime Detection Laboratory (SCDL)
Schachter, Daniel, 249
Schafe, Glenn, 264
schema, 216–20, 221
Schneck, Jerome, 116
Schneider, Sidney A., 138
Schneider Brain Wave Synchronizer, 138–39
Scientific Crime Detection Laboratory (SCDL), 48–51
scientific technique, definitions of, 49
scopolamine, 35–52, 276n6. *See also* truth serum
Scoville, William Beecher, 85–86
Scream, The, 191
Scrivenor, W. S., 37–39
Seduction of the Innocent (Wertham), 132
self, 3, 7–8, 127, 266–72; self-help and self-help literature, 189–90. *See also* past lives
sexual abuse: in the Catholic Church, 253; and false memory, 225–45, 249–55; and the feminist movement, 181, 183, 190–91; inner child, 190; and the law, 182–88, 235–36, 239–43, 250–51; and recovered memory, 180–96; and repressed memory, 225–27, 235, 239–43, 249–54; statutes of limitations, 182–88; support communities for, 188–93
Shachter, Daniel, 177
Shades of Gray, 283n47
Shanley, Paul, 253
Shattered Trust, 184
Shelley, Mary, 132
shell shock. *See* battle trauma
Shirley, Donald Lee, 154
Simon, Benjamin, 123
Singer, Margaret, 244, 248–49
60 Minutes, 226
Skeptical Inquirer, 250
Slate, 247–48
Slater, Eliot, 55

Smith, Ed, 37, 39
Smith, Frank, 300n15
Smith, Harry, 96–98, 285n34; *Heaven and Earth Magic*, 97, 98
Smith, Thomas McGinn, 193
Snowden, W. Scott, 240
social constructivism, 221–22
social engineering, 27, 28
social psychology, 132, 148
sociology of scientific knowledge, 221–23
sodium amytal, 35, 53, 54–55, 55–58, 63, 69–70, 239, 280n7
Sodium Pentothal, 35, 53, 59–73, 89–90, 126, 137
Solomon, Alfred, 120
SOUP (Survivors Oppose Ubiquitous Perps), 191
Spence, Donald, 186–87
Spiegel, David, 192
Spiegel, Herbert, 154
Spiegel, John, 59, 61, 65, 72–73, 126; *Men Under Stress*, 63, 68–69
St. Louis Post-Dispatch, 44
Stanislavski, Constantin, 99
State of Ohio v. Papp, 152
State v. Hurd, 154
statutes of limitations, 5, 182–88
Steunenberg, Frank, 21–22
Stevenson, Robert Louis, 213
Strasberg, Lee, 98–99
stream of consciousness, 78–79, 92, 97–98
stress hormones, 262, 263, 265, 266
suggestion, 5, 9, 11–14, 16, 18–19, 21, 44, 148–55
Superior Court of San Diego, Cornell v., 289n2
survivor culture, 188–92
Sutton Emergency Hospital, 55, 56, 126
synaptic account of learning, 83, 167

tachistoscope, 203, 300n15
Tales of Terror, 130, 130
tape recording, memory as a, 139, 155
Tart, Charles, 112
Taus, Nicole, 250
television, 158, 159
Terr, Lenore, 176–78, 179, 192
testimony, witness. *See* witness testimony
tetrahydrocannabinol acetate, 71
theater, 98–99
Theosophy, 112
thinking, 217
third degree, 46–47, 49
Third Degree, The (Klein), 18–20, 19

Tibet, 105
Tighe, Virginia, 103, 105–6, *107*, 114–21. *See also* Bridey Murphy affair
Time, 96, 226, 227
time: physiology of, 78–79; and witness testimony, 186. *See also* statutes of limitations
Torres Straits exhibition, 201, 210
torture, 46, 49, 51
trances, 34
Treatise on Evidence (Wigmore), 29
Treatment of War Neuroses, The, 56–57
truth: narrative *vs.* historical, 186–87. *See also* authenticity
truth serum, 33–52; and civil liberties, 48; and confessions, 40, 47–51; and corrupt institutions, 45–47; and false memory, 44; and the law, 36–43, 45–51, 71–73; and memory, 4, 43–45
Tuohy, William, 71, 72
twilight sleep, 35, 36. *See also* scopolamine
Tyson, Nancy, 186

Umbenhour, Lee, 111
unconscious memory, 34–35, 36, 43–45, 56, 91, 141–43, 271
Underwager, Ralph, 228, 229, 237–39, 299n34, 305n27

Van Derbur, Marilyn, 193
Van der Kolk, Bessel, 178

Wakefield, Hollida, 228, 238, 305n26
walking zombie syndrome, 141
war. *See* World War I; World War II
Ward, James, 200, 202, 300n6

"War of the Ghosts, The," 207–12, 214, 219
Wellesley College, 171
Wertham, Frederic: *Seduction of the Innocent,* 132
Westwood, Roberta, 120
White, Wally, 118–19
Wickersham Commission, 50–51
Wiener, Norbert, 202, 208
Wigmore, John Henry, 29–30, 275n47; *Principles of Judicial Proof,* 32; *Treatise on Evidence,* 29
Williams, Mary, 184, 187
wire recording, 91
witness testimony: and drugs, 270; and hypnosis, 129, 143–55; and psychology, 9, 24–25, 29–30, 31, 32; reliability of, 5–6, 29–30, 31, 227; and time, 186
Wolberg, Lewis, 116, 120
Woodbury, Loretta, 192, 249
word association tests, 46
World War I, and psychiatry, 54, 212–13
World War II: battle trauma, 53, 55–57, 59–69, 104; blocked memories of soldiers, 4, 53–54, 56–57, 61, 64, 72–73; and psychiatry, 53–54, 55–70, 73, 125, 126, 281n18, 281n21
Wright, M. B., 65
Wright, Teresa, 107, 286n17
Write to Heal, 191–92
Wundt, Wilhelm, 15, 200

X, Malcolm, 159

Yeames, W. F.: *Prince Arthur and Hubert,* *204,* 205

Zapruder, Abraham, 160–61, 165
zombies, 141